The Ageing Population

Polly Lama
Editor

The Ageing Population

Impact Analysis on 'Societal and Healthcare Cost'

Springer

Editor
Polly Lama
Anatomy, Sikkim Manipal Institute of
Medical Sciences
Sikkim Manipal University
Tadong, Sikkim, India

ISBN 978-981-99-5771-2 ISBN 978-981-99-5772-9 (eBook)
https://doi.org/10.1007/978-981-99-5772-9

This Springer imprint is published by the registered company Springer Nature Singapore Pte Ltd.
The registered company address is: 152 Beach Road, #21-01/04 Gateway East, Singapore 189721, Singapore

Paper in this product is recyclable.

Preface

Ageing is an inevitable, irreversible, natural process, and an integral part of the human life. Age-related changes are progressive, dynamic, interlinked, and complex as several intrinsic and extrinsic factors together influence the elderly individuals. Although ageing is not necessarily a negative process, the increasing number of human population estimated 8 billion in 2022 by the United Nations World population census report creates a distinct stratum as the 'retired ageing population'.

The significance and concept behind old age is not only the increase in number of persons in that group and their particular needs but the contrasting differences experienced by the elderlies in terms of general well-being in various countries and continent has a considerable impact in the lives, health, social and economic well-being of the aged, retired, and the vulnerable oldest old. Perhaps, some of the common and greatest risk factors that impacts the retired are the losses they face due to aged-related physical constraints limiting earnings, decline in socio-economic viability, and health-related costs mainly due to recurring chronic illness, injuries, and diseases. These factors influence the sustainable development goal plans adopted by the various organisations and countries in the World to end hunger, poverty, social insecurities and to promote good health, economic growth, and prosperous communities. Thus, the main **'aim and purpose'** behind this book is to provide a comprehensive overview on the World ageing population and to extensively discuss the impact and significance of the increasing ageing population on public programmes such as pension plans, social welfare aids, healthcare cost, and labour compensation. This books also include chapters focusing on demographic distribution of age-related diseases and disabilities that increases the economic burden towards long-term care and social psychological reforms. Chapters also focus on understanding the current research and development prospect adopted to ease the ageing process, especially in the European region where the rapidly ageing group is gradually 'shrinking' which is in contrast to many South Asian, African, and North American countries where the aged group and the oldest old is 'expanding' thus 'what does healthy ageing mean' is discussed in detail in this book.

The knowledge base behind this book is augmented by published analytical research studies on population ageing and its distribution pattern based on socio-economic and healthcare context, and the strategies adopted to save cost, time, and efforts required to devise policies, plans, and principles in sustaining a prosperous ageing population with minimum liabilities. The comparative approach behind differences in ageing between continent and countries in this book provide a fuller and deeper understanding on ageing trends and challenges worldwide to present the most essential and cogent strategies for formulating positive, prosperous, and sustainable ageing policies equipped with developing newer technology-enabled approaches to serve the ageing community.

Chapter 1 provides a thorough explanation on the distributions and diversities behind the ageing population across the major countries and continents of this World. Globally accepted term and terminologies are used to measure or define the similarities and differences behind the growing ageing population. Various essential factor associated with ageing, including socio-economic policies, general well-being, and healthcare issues are described continent wise to allow the readers to get a clear overview about the factors that affects an ageing individual, including the country and the continent.

Chapter 2 presents the impact of disability and diseases amongst the aged with recent research and therapeutic advancements. For example, osteoarthritis as a major musculoskeletal disease and disability of the elderly worldwide, cardiovascular aliments as the major diseases leading to mortality, the impact of COVID-19 on the aged, and the most recommended treatments used to alleviate such concerns.

Chapter 3 describes about the feasibility of psychosocial interventions on the quality of life for the aged elderly and examines evidence from diverse field of research on care planning, implementation programme, and outcome of holistic care for the vulnerable elderly guided by inputs from the stakeholders involved in this task.

Chapter 4 describes and analyses the cost involved with treating diseases and disabilities that affects the elderly and examines the influences on a demographic region. Comparisons are drawn between developed, developing, and under-developed countries and continents of the World. This chapter also highlights the challenges, opportunities, and differences in the ageing economy with relevance to expenditure incurred by the healthcare system to present the most cost-effective healthcare practices.

Chapter 5 focuses on priorities for future investments that can be adopted by various countries and regions to sustain the ageing group. The best practices, policies, and prospects with credible positive evidence are highlighted for further discussions to suggest the most viable strategies.

Chapter 6 delivers a conclusive overview on the dynamics behind the World ageing population with some critical questions for the hour.

Tadong, Sikkim, India Polly Lama

Acknowledgements

As human beings, we can't ignore how our lives can depend upon several factors that surround us, and variabilities are thus understandable. All of us, undoubtedly, would like to live a productive and healthy life with minimum shortcomings, yet some of us can't, as our health, well-being, and employability depend on social, economic, environmental, and political factors. The polices as adopted by the respective countries, worldwide therefore will have a significant impact on our lives and the prospect of healthy ageing. Thus, this book acknowledges the researchers, policy makers, and experts who have contributed to rationalising the ageing process across the World, and to the readers of this book who can think and plan about how can we better shape and serve the ageing community based on the existing resources, thus the editor of this book also acknowledge the Sustainable Development Goal plans as adopted by United Nations and followed by the World.

Contents

Contributors

Sonam Choden Bhutia Department of Biochemistry, Sikkim Manipal Institute of Medical Sciences, Sikkim Manipal University, Gangtok, Sikkim, India

Chamma Gupta Department of Biotechnology, Sikkim Manipal Institute of Medical Sciences, Sikkim Manipal University, Gangtok, Sikkim, India

Bidita Khandelwal Directorate of Research, Sikkim Manipal Institute of Medical Sciences, Sikkim Manipal University, Gangtok, Sikkim, India

Polly Lama Sikkim Manipal Institute of Medical Sciences, Sikkim Manipal University, Gangtok, Sikkim, India

Harshvardhan Sampath Psychiatry, Central Referral Hospital, Sikkim Manipal Institute of Medical Sciences, Sikkim Manipal University, Gangtok, Sikkim, India

Geeta Soohinda Psychiatry, Central Referral Hospital, Sikkim Manipal Institute of Medical Sciences, Sikkim Manipal University, Gangtok, Sikkim, India

Binod Kr Tamang Department of Anatomy, Sikkim Manipal Institute of Medical Sciences, Sikkim Manipal University, Gangtok, Sikkim, India

Abbreviations

AD	Alzheimer's Disease
AIDS	Acquired Immunodeficiency Virus
CVD	Cardiovascular Diseases
COPD	Chronic Obstructive Pulmonary Disease
COVID-19	Corona Virus Diseases of 2019
DALY	Disability Adjusted Life Years
DM	Diabetes Mellitus
EAN	European Ageing Network
EPF	Employee Provident Fund
FAD	Familial Alzheimer's Disease
GAN	Global Ageing Network
GDP	Gross Domestic Product
HIC	High Income Countries
HIV	Human Immunodeficiency Virus
IMF	International Monetary Funds
LIC	Low Income Countries
LDCs	Least Developed Countries
MIC	Middle Income Countries
NCDs	Noncommunicable Diseases
NGO	Non-Government Organisations
NHC	National Health Commission
OA	Osteoarthritis
OECD	Organisation for Economic Co-operation and Development
PD	Parkinson's Disease
RM	Malaysian Ringgit
SAD	Sporadic Alzheimer's Disease
SACRED	Senior Citizens for Re-Employment in Dignity
TFR	Total Fertility Rate
UAE	United Arab Emirates
UN	United Nations

US	United States
UK	United Kingdom
VaD	Vascular Dementia
WHO	World Health Organization

List of Figures

Chapter 1
Continent Wise Intersectional Analysis on Ageing

Polly Lama

Abstract Persistent advancement in healthcare has increased the average life expectancy globally in the last few decades, increasing the proportion of the aged, and the oldest old. Published past findings indicate significant variabilities exist between countries and continent with relevance to societal well-being, pension policies, and labour compensation. Thus, the prospect of ageing gracefully especially in the case of he old unable to work further pose as an additional burden to the ageing society and the World. The aged will live longer, and even though the current advocated healthcare practices has added years to life, the expenditure incurred, and difference in the quality for care offered questions the credence of the current system. Careful planning to formulate the best-aged care policies and plans may still be a pending need to objectively justify the factors addressing the inconsistencies faced by the aged and the retired.

Keywords Ageing · Healthcare · Expenditure · Well-being · Care

1.1 Introduction: The Ageing Community

Globally, individuals aged 60–65 years and above are considered fundamentally as the retired elderly population [1]. The average statutory retirement age of 38 countries in 2018, as mapped by the Organisation for Economic Co-operation and Development (OECD) was 64.2 years for males and 63.5 years for females [2]. OECD defines the normal retirement age in a given country as the 'eligibility of all reform schemes combined, based on a full career after labour market entry at an approximate age of 22 years. Government and private sector policymakers normally consider the fiscal cost of ageing, life expectancy, nature of the profession, supply of labour force, healthcare infrastructure, and dynamics of the demography whilst considering the retirement ages [3]. Sixty-five years is the maximum accepted age

P. Lama (✉)
Sikkim Manipal Institute of Medical Sciences, Sikkim Manipal University, Gangtok, Sikkim, India
e-mail: pollylama@smims.smu.edu.in

P. Lama (ed.), *The Ageing Population*,
https://doi.org/10.1007/978-981-99-5772-9_1

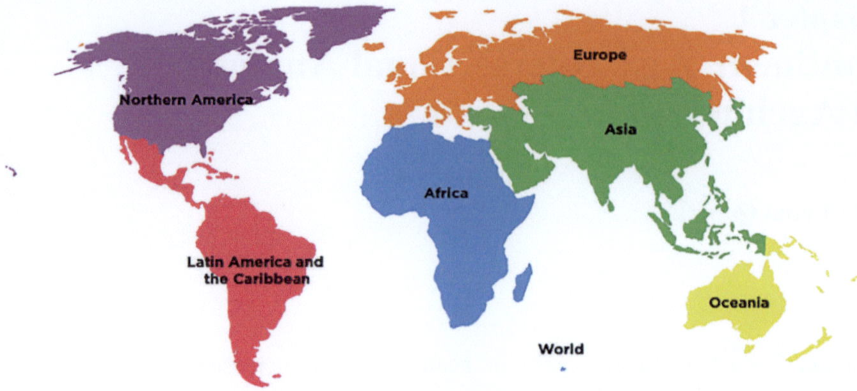

(Numbers in millions)

Region	Total population		Population aged 60 and older			
			2020		2050	
	2020	2050	Number	Percent	Number	Percent
World	**7,684.3**	**9,665.3**	**1,045.4**	**13.6**	**2,092.2**	**21.6**
Africa	1,339.5	2,533.6	74.4	5.6	235.1	9.3
Asia	4,539.6	5,188.0	603.2	13.3	1,292.2	24.9
Europe	749.3	716.4	191.1	25.5	246.7	34.4
Latin America and the Caribbean	644.3	742.6	82.6	12.8	183.4	24.7
Northern America	370.5	432.1	86.8	23.4	122.2	28.3
Oceania	41.1	52.6	7.3	17.9	12.6	24.0

Fig. 1.1 The World Regions and the total population for those aged 60 years and above in 2020 and projected 2050, data adapted from US Census Bureau, International population database, 2022 [8]

to retire in seven continents of the world, where a retiring person may be entitled to receive superannuation or benefits as fixed by the state or country-specific pension plan/policy or scheme, and which is on an average almost the half of the salaries earned by the retiring individual, especially amongst the developing countries [4]. A major global concern of the ageing community regarding pensions is the increase in life expectancy due to advancements in medical technologies which necessarily does not improve the quality of life lived by the aged and retired [5], thus, an argument is posed against increasing the current age of retirement [6] as an agenda of the twenty-first century sustainable development plan [7].

In 2022, the United Nations population bureau reported a total of 8 billion global population, and with an approximate 4 billion, or 55% of the World population, China and India with 2 billion each forms the two most populous countries in the Eastern and Southern Asian continent, and the World (Fig. 1.1). The African continent has a total population of 1.3 billion, whilst Europe and North American counterparts are at 1 billion each, followed by Latin America and the Caribbean at 658 million and the Oceania continent at 44 million [8].

The 2022 World population prospect as reported by the United Nations confirmed for the first time as individuals aged 65 years and over outnumbered children under the age of 5 years in 2018, and there are approximately 771 million people aged 65 years or over globally, which is three times more than in 1980s. The report further predicts the retired population will reach a mark of 994 million by 2030 and 2 billion approximately by 2050 (Fig. 1.1). Thus, by 2050 there will be twice as many persons aged 65 years or older than children under 5 years in the World, and the number of individuals aged 65 years or over will be almost the same as the number of children below 12 years. The forecasted population growth report for 2050 therefore suggests that there is an impending need to reform or develop new old age policies focusing on economically viable approaches such as re-employment opportunities post-retirement to create social security, or, by increasing the average age of retirement to 67 years and improving the long-term geriatric healthcare services for the oldest old aged 75 and above, however, such reform measures needs to be calibrated at a practical level to ensure sustainability for the growing ageing population [9].

Therefore, to understand, and to analyse the ageing community worldwide and their current issues, the next subsequent sections in this chapter will explain and examine continent-wise similarities and disparities within the ageing community, to highlight and to outline the overall holistic factors that significantly affect the lives of the elderly.

1.2 Ageing Trends: India and South Asia

The technical population projection group of India in 2021 reported a total of 138 million elderlies above the age of 60 years, 67 million were males and 71 million as females [10, 11]. In Indian communities, historically cultural and social heritage system has placed the elderly as the revered who are usually sheltered within the joint family system [12], however, these older practices are disappearing rapidly due to changing economic trends, modern-day nuclear family settlement practices, natural disasters, and health-associated costs [12], thus the older people are left very much on their own, especially in the remote rural parts of India (Fig. 1.2).

In contrast, many affluent urban societies due to their own private earnings and assets investment plans remain unaffected as they can afford benefits such as luxury retirement schemes, aesthetic housing, and five-star healthcare facilities with ease [13]. The poorer section of the population which forms the larger percentage of the ageing population in India thus suffers considerably, as the spectrum of paid facilities on par with the developed countries for the aged in India are left as much to be desired by the poorer community. To overcome, such complexities for the elderlies below the poverty line, classified as the retired unorganised job group, the Government of India has introduced many social welfare schemes since 2010 such as the National social assistance programme, Indra Gandhi National old age pension scheme, Atal pension plan in 2015, along with the establishment of the National Centre for ageing in 2017 for few urban metropolitan cities, and even more recent

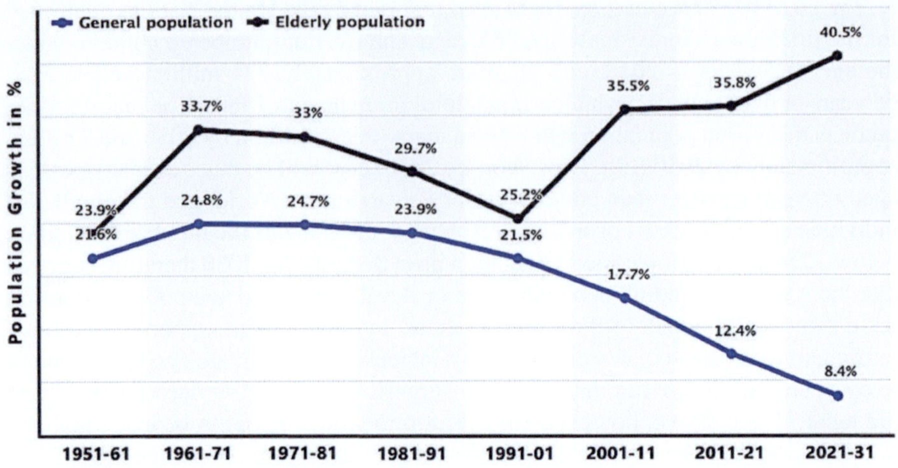

Fig. 1.2 Population growth in India over the decades, Technical population projection database of India, 2021 [10, 11]

initiatives and plan such as the senior citizen re-employment in dignity (SCARED) scheme in 2021 [14, 15]. Similarly many private Non-Government Organisations (NGOs) facilitate housing, re-employment, and 2–3 meals a day programme through private and community-based charities [16]. Many primary, secondary, and tertiary care geriatric services offer free healthcare programmes which are operational nation-wide since 2016. The establishment of 19 regional geriatric centres in tertiary care hospital services under the National programme for the healthcare of the elderly is another such care prospect [17]. India also plans to develop several new National Centres for Ageing, and three of them located in the state of Delhi, Madras, Chandigarh, nearing their full completion by 2024. Further, in 2018 the ministry of health and family welfare of India introduced a health coverage plan of up to 100 million for persons belonging to poor vulnerable families and the aged.

Apart from India, other countries that form the Southern Asian continent are Afghanistan, Bangladesh, Bhutan, Iran (Islamic Republic of), Maldives, Nepal, Pakistan, and Sri Lanka with a total approximate population of 1.6 billion (excluding India), and in contrast to India, the rest of Southern Asian countries project a lower proportion of the older people within their own regional domain at 10.9% [18]. Even though, many studies claim and report large number of total populations in countries such as India, Pakistan, Bangladesh, and Sri Lanka, this does not directly mean more number of elderlies because the mortality rates[1] are high amongst the older adults in these countries with an approximate average life expectancy of 70.5 years [19]. Most of the Southern Asian countries also experience harsh and unfavourable

[1]Mortality rates: Mortality is typically represented as a rate of death per 1000 individuals in one year, also called the death rate. For calculation purposes, the mortality rate is deduced by dividing the number of deaths in a given time for a given population by the total population.

demographic locations further attenuated by extreme poverty, shortage of adequate food supplies, lack of accessible medical facilities, insurgence of healthcare pandemic such as COVID-19, especially during 2020, internal political disruptions, terrorism, war, as some of the main factors that rapidly contributed towards a sharp decline in the fertility rates [20].

Pakistan, India, and Afghanistan are the three South Asian countries with the highest death rates due to chronic illness, diseases, war, lack of adequate medical facilities and pollution [21]. Pakistan ranks amongst the top two countries globally in terms of infant mortality rates, whereas Maldives, Bhutan, and Iran in contrast have the lowest death rates amongst all the other South Asian countries.

Thus, to summarize, many South Asian countries, even after the rapid twenty-first century advances in digital technology sector still face poverty, and hunger issues as the main problem centred around slow economic growth that may rapidly compete with the rising population of the elderly, and which to an extent has shifted the focus of many South Asian countries from the ageing group. Another additional geographical disadvantage of the South Asian countries is the lack of adequate arable agricultural land for food production in comparison to the total population [22]. The number of people per square mile of arable land, often called as the 'physiologic density' tend to be high in South Asia due to large areas of land being occupied by deserts, mountains, or as islands [23]. Further, catastrophe such as wars, natural disasters, and incidence of diseases has caused many more people to die than expected under normal circumstances, often termed as excess mortality rates,[2] plagued further by frequent terrorist combats, food shortages, and unmet health needs largely attributable to the recurrent destruction of essential infrastructure and saturation of hospitals particularly after 2020 [24].

Old age financial security and well-justified pension policies are still at an incipient stage thus lacking in mass administration, and approximately 70% of the older group are still not covered with any type of pension schemes in the South Asian countries. One of the main reasons behind non-uniform availability of standard pension plan is lack of formal education due to poverty, and vast majority of the working population still being employed within the agricultural employment sector [25].

Thus, to be 'aged and retired' in India and the South Asian continent has a variable meaning, and differs between countries, states, urban and rural regions as differences in private earnings, and between employed, non-employed, self-employed, agricultural, and manual labour.

[2]Excess mortality rates deaths per 100,000, and the number of reported all-cause of deaths is higher than expected deaths.

1.3 Ageing Trends: China and East Asia

With the fastest-growing ageing community, China is one of the most populous Eastern Asian country and presents a total of 267 million persons aged 60 years or above [26]. The release of the seventh national population census in 2021 (Fig. 1.3) reported that China has attained a moderately prosperous society with the eradication of abject poverty [27], and 40% of Chinese people earn an average of 1000 Chinese Yuan currency per month which is approximately equal to around US $ 150 [28]. The Chinese republic has also launched many active policies such as the national initiative for ageing population, where the nationwide focus has been towards planning, developing, and implementing social reforms, better healthcare facilities, and creating adequate employment opportunities for the aged and retired [29, 30]. Chinese organisations confirm many ageing-related welfare measures are adopted by the country that works in tandem with the reformed social policies to make ageing population an integral part of economic growth, well-being, and prosperity [31].

New policies laid by the National Health Commission (NHC) and Ministry of Civil Affairs in China aim to create ageing friendly cities and communities where older people continue to make meaningful contributions to society by exploring innovative approaches towards integrated care, such as the establishment of an internet based medical information portal for major and minor illness [32]. In 2019, China reported 75% of the elderly aged 60 years and above suffer from non-communicable diseases such as cardiovascular complications, diabetes, and hypertension [33]. Thus, China has been exploring innovative approaches to treat

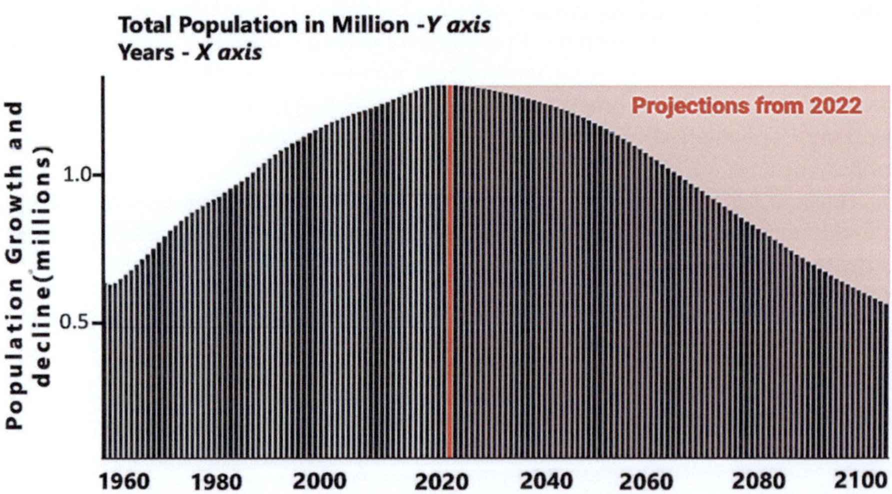

Fig. 1.3 Chinese population projection survey indicates a decline in the total population post-COVID-19 pandemic, as per the report released by the Shanghai Academy of Social Sciences, 2021 [32, 34]

chronic diseases to foster healthy ageing prospects. This has led to tripling of heath expenditure since aged-related health reforms began in 2010 form 481.6 billion Chinese Yuan to 1520.6 billion CNY in 2017.

In 2022, the World Economic Forum for the Chinese ageing gap study reported for the first time that persistent challenges are encountered in sustaining re-employment opportunities for the elderly and this growth represent as a major economic concern, which includes upgrading the current healthcare services for individuals aged 70 years and above [34].

Apart from China, other Eastern Asian countries comprise of Hong Kong, Macau, the Democratic People's Republic of Korea, North Korea, Japan, Taiwan, and Mongolia. The total population of East Asian countries was at 2.6 billion in 2022 forming 21% of the total world population, and China, as described earlier, alone has the largest total population count at 2 billion approximately. These figures clearly indicate that the rest of the East Asian countries, other than China, have a relatively lower number of the total population, however, this also does not directly mean a lower number of elderlies in East Asia. The ages between 60 and 65 years in all East Asian countries are growing rapidly at a rate of 30% in comparison to the rest of the world at 20% [18]. The main reason behind such drastic ageing growth percentages is declining fertility and increasing life expectancy since 1950 [8]. Thus increasing number of elderlies in East Asia with declining fertility rates confirm a shrinking working age in the East asian region, which is significantly higher than the rest of the world. This peculiar ageing trend is often referred to as 'compressed modernity' favouring economic growth and socio-political transformations, but it has weakened traditional values of filial piety that expect adult children (sons in particular) to care for their ageing parents. Traditionally, in many East Asian countries, a strong family-dependent welfare scheme for older adults usually act to provide as a adequate public safety nets to meet the needs of the older population [26]. Thus, the increasing number of elderly population has begun to exert enormous pressure on the working class families, healthcare, labour markets, and various social reform [35, 36]. It is against this backdrop the ageing group is most impacted in the East Asian countries [37, 38].

Japan, in particular, and amongst all the other East Asian countries and the World, has the highest numbers of the oldest population [39]. One in four people in Japan, are at least 65 years old [40]. Low fertility rates and a smaller number of working-age adults together contribute towards declining the total number of productive Japanese population and, by 2050, Japan is projected to shrink by 20 million people [8, 35]. These population estimate reports suggest that the older ageing population will rise significantly for the next three decades, followed by a sharp downfall in the total population thereafter. Lower birth rates also suggest that there will be fewer females in the child bearing age, which increases the old age dependency ratio[3] in Japan and it has been estimated that there is one person over 65 years of age for every two people between 20 and 64 years [41].

Such odd pattern disrupts economic growth as it shrinks the labour from productivity towards increased consumption of resources. Other East Asian countries such as Korea, Macao, Taiwan, and Hong Kong have begun to report similar trends as lower fertility rates and higher life expectancies including China which will impact the population dynamics of East Asian countries [35, 42].

Addressing the potential future challenges, large population-based surveys with rigorous methodology suggest disparities in socio-economic conditions that determines the quality of remaining life lived amongst the East Asian elderlies further influencing cognitive functioning and transitions out of the labour market as the main indicators diminishing the value of life in the older population. Many studies also demonstrate family relations and networks shape an individual's final year of life and their children's perception of ageing [10].

1.4 Ageing Trends: South-Eastern Asia

The Southeastern Asian continent had a total population of 674 million in 2022 [43, 44]. Indonesia as a constituent country of Southeast Asia has the largest population of 278 million [45]. Indonesia also has an approximate total of 16,000 islands including Philippines which is currently the fourth most populous country in the world following India and China [8, 46]. Southeast Asian continent accounts for nearly 12% of the World's total population. However, much like Southern and East Asian continent, there are considerable differences in ageing trends between the geographically dispersed territories. Indonesia accounts for two-fifths of the regional total population as aged and retired which is in contrast to Brunei's contributing only a fraction of that with a mere 0.449 million in 2022. The other mainland states comprising of Malaysia, Myanmar, Singapore, Cambodia, Laos, Timor-Leste, Vietnam, and Thailand account for about half of the remaining older population.

Typically, 50% of the Southeastern Asian population relies on agriculture as the main source of employment and economic growth [47], thus agricultural output has increased significantly since the 1970s, however, there are considerable variations in economic prosperity between the South Asian countries, with the greatest economic gains reported in Singapore, Malaysia, and Thailand and little or no increase in Cambodia, Laos, and Vietnam due to crop failures and internal political strife. Hunger and malnutrition are the main problems in many South Asian countries [48], and are especially serious in Cambodia. Much like in East Asia, poverty is reported more within the rural population and is related to limited land access, and further the landless person experiences greater poverty and poorer health [49] Land

[3] Old-age dependency ratio compares different population groups dividing those considered 'dependent' over those considered 'productive'.

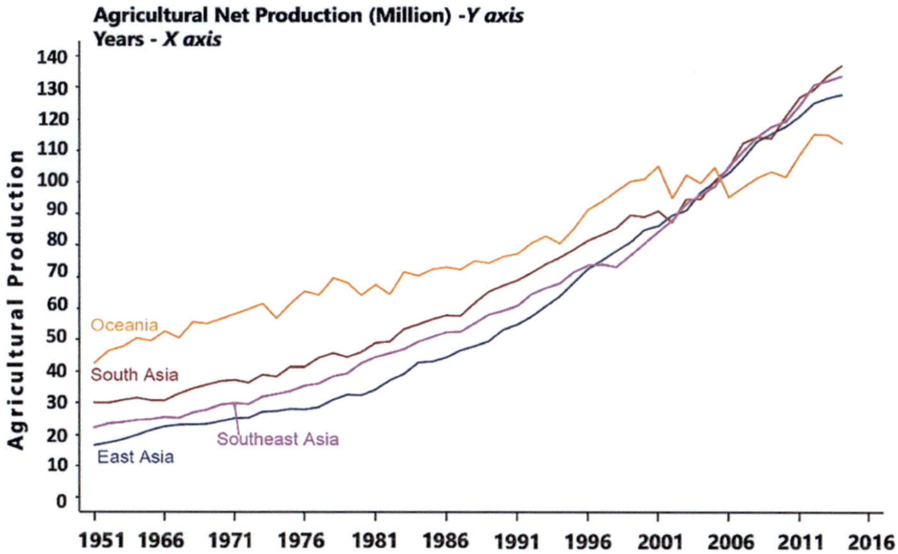

Fig. 1.4 Food production indices in Southeast Asia, a major source of non-pensioned employment. United Nations, Food Agricultural Organisation [10]

access problem is critical in the Philippines with the lowest available area of arable land [50]. Singapore, is a notable exception amongst Southeastern countries, as the economy has boomed since the advent of Industrialisation [51]. Southeast Asian countries with largest proportion of export trade are Singapore, Malaysia, and Brunei, thus there is a considerable country-wise disparity in the economic development within the same continent, which influences the care facility and socio-economic opportunity received by the ageing population.

Sixty percent of the older populations in Southeast Asia rely on agricultural wholesale retail trade and manufacturing industries for employment and financial security (Fig. 1.4). Agriculture as the main source of occupation clearly suggest that country and state-specific pension plans cover only a small minority of the older adults [35, 52]. A study on the Government operated Malaysian Employee Provident Fund (EPF) scheme revealed that 21.5% of the members aged 51–55 years had zero balance in there saving plans and 46.2% had less than RM50,000 [52, 53], which is largely due to rapid economic, cultural, and social changes and challenges eroding the family care values to support the old age individuals, resulting in a smaller pool of potential caregivers within the family. The Employee Provident Fund (EPF) and Pension (for civil servants) are inadequate to meet the financial needs of older people [54] in many Southeast Asian countries. Therefore, older people continue working even after the age of 65 years to support themselves and their dependents.

The most important concern for the social security and/or pension systems in many parts of Asia is low coverage for the labour force regardless of the workforce involved with agricultural output. Pension coverage is especially low in India, Pakistan, and Indonesia almost below 20%, which is in contrast to 79.9% in South

Korea and 95.4% in Japan [4]. These differences clearly indicate that old age social security coverage is related to national prosperity, and can be extended to non-employed labour forces or those involved with manual labour.

1.5 Ageing Trends: Central Asia

The Central Asian continent comprises five republics, and they are Kazakhstan, Kyrgyzstan, Tajikistan, Turkmenistan, and Uzbekistan with a total population projected as 77 million [8]. The Central Asian region has the most unusual geographical location with landlocked borders, mountainous terrain, and Eurasian steppes, thus agricultural productivity is largely unfavourable due to limited availability of water and fertile land resources [55]. Therefore, a majority of the population including the elderly earn their living by herding livestock in rural areas and through industrial activities in urban cities (Fig. 1.5).

The total number of retired individuals residing in Central Asian countries was reported as approximately 13 million in 2022, forming 10.7% of the total population [56]. Many Central Asian countries have been considering raising the retirement ages for the working sector, or to adopt re-employment opportunities, especially within the urban areas of Uzbekistan. The two primary reasons for the increasing proportions of elderlies in Central Asia are attributable to significant progress in

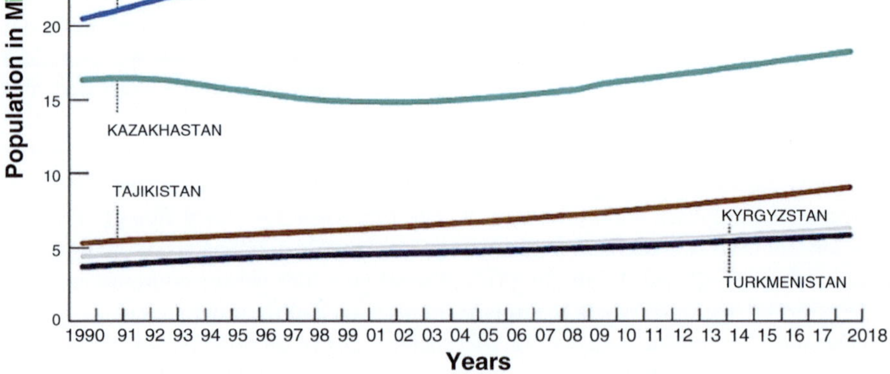

Fig. 1.5 Increasing population of Central Asia is documented with an increase in individual account holdings, as reported by the World Bank Reports, 2022 [41, 45, 95]

healthcare for older person increasing the average life expectancy. The populations of Central Asia has doubled since 1990 and is expected to increase to 36.9% by 2050 [8], thereby contributing towards crossing the one billion mark of the global elderlies, UN, 2022. Current and past 10 year population trends indicate Uzbekistan is the most populous urban state [57] with a total population of 34 million people, followed by Kazakhstan (19 million), Tajikistan (9.5 million), Kyrgyzstan (7.5 million), and Turkmenistan (6 million). These current and past figures clearly suggest that Central Asia is a relatively fast-growing population region much like the rest of the Asian continent, however, the main difference is the higher younger median age group in Central Asia, thus, a larger share of working age population that currently aid the economic productivity, but this set-up will eventually and perphaps peak to its maximum efficacy and began to decline in the future as the old-age dependency ratio is set to increase in another three decades [58]. Fifty-two percent of the Central Asian population depend on agriculture as their main source of income, and the remaining 48% rely on jobs within the industrial sector [59]. Thus, a major portion of the population involved with agricultural as employment prospect would mean old age security coverage and well-developed pension policies needs to be addressed and which is mostly non-existent in Central Asian countries, and the elderly remain in labour force longer because when they stop working they tend to be fully reliant on their families for financial support [60, 61]. Central Asian countries thus need to prepare systemically for sustaining the uprising demands of the older population with relevance to economic stability, food, and health security considerably.

Traditionally, the oldest old in Central Asia tend to live with extended families and are cared by them who provide a means for subsistence and financial support. This type of arrangement is mostly prevalent in densely populated rural areas of economically weak countries such as Kyrgyzstan, Tajikistan, and Uzbekistan. To lessen the cost for care of older adults, after 2014, the working age group in Central Asian countries are encouraged to invest in contributory pension funds [4, 62], however, the funds received are at a bare minimum, and Central Asian Governing bodies are still yet to develop a robust welfare system. Low literacy rates are another factor depriving older adults of re-employment opportunities, and there is still no documented provision for encouraging literacy amongst the older aged.

1.6 Ageing Trends: Western Asia

With an approximate total of 288 million population in 2022, the Western Asian continent comprises Armenia, Azerbaijan, Bahrain, Cyprus, Georgia, Iraq, Israel, Jordan, Kuwait, Lebanon, Oman, Qatar, Saudi Arabia, State of Palestine, Syrian Arab Republic, Turkey, United Arab Emirates, and Yemen. The two most populous countries of Western Asia are Turkey (84 million) and Iraq (40 million) whilst Bahrain (17 million) and Cyprus (12 million) are the least populated areas [8, 63]. The overall urbanisation rate in 2022 was at 72%, which is high in

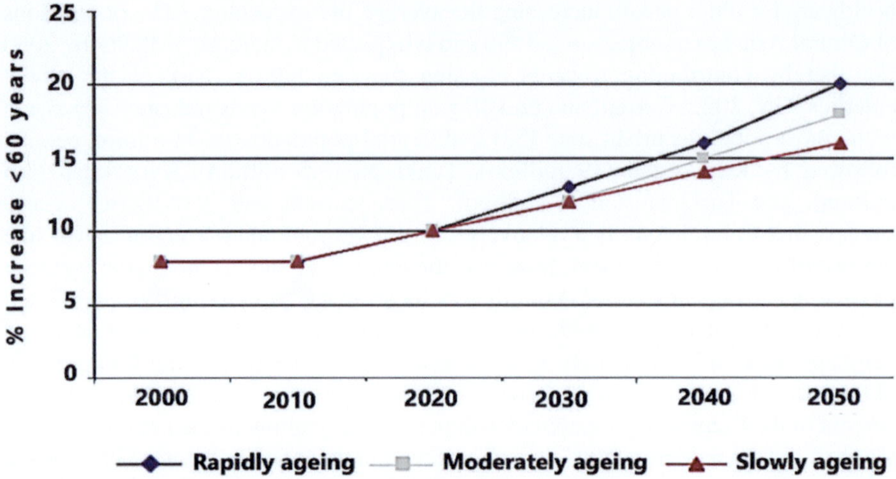

Fig. 1.6 Percentage increase in older adults of rapidly, moderately, and slowly ageing Western Asian countries, data extracted from UN population database, 2022 [8, 67]

comparison to other Asian regions, and the majority of the people live in the cities largely due to a general shift from an agriculture-based economy to the rapid development of mass industry, technology, constructions, and IT services contributing towards economic growth [64]. In comparison and contrast to the East and South Asian countries, the overall old-age dependency ratio of Western Asia is at 8.7% [65] comparable to Latin American and European countries, therefore Western Asia is still experiencing a slow to moderate pace of ageing, but a few countries such as Turkey and Lebanon have more than 10% of their populations aged 60 years and above. Bahrain, Saudi Arabia, Oman, United Arab Emirates, Armenia, and Azerbaijan are expected to join this group by 2030, and the State of Palestine, Egypt, Jordan, Syrian Arab Republic, and Qatar by 2050 [8]. Therefore, Western Asia is often grouped as rapidly, moderately, and slowly ageing countries (Fig. 1.6).

'Slowly ageing' and developing countries of Western Asia are Iraq, Yemen, Syrian Arab Republic, and the State of Palestine. Lower proportion of elderly population in these countries is due to slow socio-economic development, high mortality rates, poor healthcare facilities, internal political conflicts, and intermittent wars compromising economic development. Jordan, Egypt, Kuwait, Oman, Qatar, Israel, Iraq, Cyprus, and Saudi Arabia are considered as 'moderately ageing', whilst Bahrain, the United Arab Emirates, Georgia, Armenia, Azerbaijan, Turkey, and Lebanon are considered as 'rapidly ageing'.

The reason behind the difference in ageing pattern amongst the Western Asian countries is due to disparities in socio-economic progression and adaptability towards advancement in technology [66]. Rapidly ageing countries also had earlier decades of high-fertility rates with a decline in infant mortality rates, thus increase in younger age groups, and life expectancy with a consequent and consistent increase in the population aged above 60 years [35]. Moderately ageing countries even with occasionally high infant mortality rates, still have a considerably large proportion of working-age groups, to progress as increased elderlies in the future. In contrast, slowly ageing countries have a typical feature comprising of fewer working-age adult population, contributing towards low fertility rates, and abject poverty making the sustenance of older ages less viable, therefore the average life expectancy differs in slowly ageing, moderate, and rapidly ageing Western Asian countries (Fig. 1.6).

Approximately 65% of the Western Asian countries still lack in formulating a cohesive standard guidelines on old age pension plan policies, especially, in the hostile war-inflicted countries, and family support is the most widespread system of care for the elderlies above 65 years of age and the vulnerable oldest old [62], which is similar to the other Asian continent. In the absence of a well-defined old-age financial schemes, some of these countries have adapted ageing-friendly policies and programmes. Jordan, Lebanon, Oman, and the State of Palestine provide provision for free public transport, special elevators, and mobility ramps. Oman has developed programmes that incentivise individuals who care for older adults, including in-home care with volunteer elderly sitters schemes and even surrogate families schemes. Meals on wheels are offered by Lebanon, State of Palestine, and Sudan, and elderly mobile care medical units are offered by Kuwait. Oman and the State of Palestine also report home-based care services either through ministerial or civil society programmes. Such services are also being developed in Lebanon, a rapidly ageing country [67].

The United Arab Emirates (UAE), in 2020 population census survey reported 1.6% of the older adults above 65 years of age were dependent on 100 working-aged people, however, by 2050, this number is estimated to rise to 18.5% [45] a 12-fold increase is therefore expected to create significant financial, social, and medical pressures [8] on the families and the Government of the UAE. The current arrangements for aged care are limited to a few local authorities and for 'emirates citizens only', and medical services for other older immigrant adults are typically provided by the local hospitals. Thus, UAE is expected to experience constraints to accommodate the rising number of older people in the future. Lack of advanced medical infrastructure, changing social dynamics, and decline in care offered by extended family model due to growing trend of downsizing households, further compounds the ageing care issue for elderlies in Emirates [67], who are often cared for at home by their own families or with the help of 'non-trained' housemaids. As of 2017, only two residential aged care facilities are exclusively dedicated to the elderly, thus the senior Emirates are forced to rely on a limited range of services provided by local hospitals, with only a few specialised medical staff in geriatrics care services [68, 69].

Thus, due to the current scarcity of senior care facilities and lack of research on ageing-related issues, the Sheikh Saud bin Saqr Al Qasimi Foundation is keen on developing new age policies, and have taken-up a research based study to design a system as conducted by the Australian Government's Royal Commission for Aged Care Quality and Safety, as Australia ranks as the best in the World in terms of human development and life indexed quality [70].

1.7 Ageing Trends: African Continent

Total population of the African continent was 1.3 billion in 2022, and Africa is considered as the world's youngest region in terms of population distribution pattern presenting a significantly modest ageing process in comparison to the Asian continent [71]. 2022 UN census bureau reported 74.4 million individuals aged 60 years and above form the retired ageing population, and the lower proportion of the elderly population is additionally confirmed by the African Union rights of the older person entitlement act [72]. Further, the 2022 UN report estimated that the elderly population may remain low over the coming decades. The key factors contributing towards fewer aged and retired individuals are directly linked to increased child and adult mortality rates [73], declining fertility [74], with little or no emphasis on the problems and well-being of the older ageing population, lack of quality healthcare facilities, literacy rates, economic and social welfare measures, and reforms, thus, the total older African population is set to remain low over the coming three decades (Fig. 1.7).

Nine in ten older African adults have no formal education, and older women in rural areas remain unemployed throughout their lives preforming unpaid domestic work [71, 75], therefore, a majority of ageing adults live in large, multigenerational households, and a significant share also live alone or in skipped-generation arrangements. According to United Nations, 'skipped-generation households' are those where the grandparents live with their grandchildren without the parents of the children in the household due to either labour-related migration, death, or frailty of

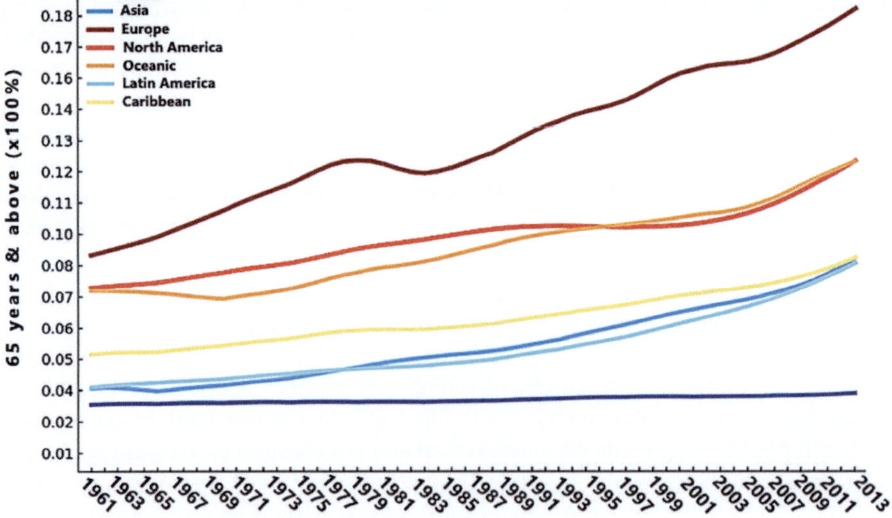

Fig. 1.7 Ageing population of the world, 65 years and above comparing the African continent with the other regions, UN population database [71]

the middle age generation especially following HIV/AIDS infection, or fostering children born out of wedlock [71]. Thus, the care burden for the dependent children and absence of a working-age adult imposes substantial pressure on older persons in the skipped-generation households, as they provide financial support to their grandchildren [76], which is a unique system present only in the African region.

For the purpose of description and geographically, Africa comprises two main divisions—the Northern Africa and the Sub-Saharan African (SSA) regions. SSA is further classified as Eastern, Middle, Southern, and Western Sub-Saharan Africa.

1.7.1 Northern Africa

Northern Africa comprises of Morocco, Algeria, Tunisia, Libya, Egypt, Sudan, and Western Sahara, including the territories disputed between Morocco and the Sahrawi Republic. With a total population of 257 million, the highest percentages of the aged individuals reside in Tunisia at 13.5%, Morocco at 11.1%, and Algeria at 9.4%, which is in contrast to the other African countries at an approximately 7% as the growing ageing population [67]. With relevance to the Asian countries, successive falls in fertility or sustained low fertility levels changed the age structure and increased the number of aged individuals, forming an ageing society, but, in Northern Africa and the Sub-Sharan region, declining fertility follows increasing mortality rates, and the 2022 UN and US census bureau report confirms, all the world regions excepting Africa has a total fertility rate[4] (TFR) below the global average of 2.1 [77, 78].

Egypt is the most populous country in the Northern Africa, with 50% of the population residing in the borders, and an approximate total of 104 million. Algeria, Morocco, and Tunisia are the other countries with approximately 45 million, 37 million, and 12 million population, respectively [8]. Since 1994, Tunisia has formed an elderly protection law that reinforces family as a primary caregiver, as the duty of the State and by the age care society, to offer protection for the oldest old unable to work [79]. Culturally, placing older individuals in care homes is not preferred generally in Africa and is considered as the last resort often opted when other channels to protect the ageing individuals within the immediate family environment fail [80]. In Egypt, publicly owned and civil society-operated old age homes are some of the facilities accessible to elderlies with low incomes.

These old age homes are operated by organisations with 100% Government assisted subsidies, whilst publicly owned care homes common in Tunisia normally meet 90% of their expenses through Government funds.

A significant issue with re-employment and ageing in Africa currrently is the lack of formal education and poor literacy rates [81]. Eighty-five percent of the employed

[4]Total fertility rate according to WHO, sum of age-specific fertility rates (usually referring to women aged 15–49 years).

older adults are thus paid informally by income generated through agriculture or manual labour work [82]. Statutory retirement ages and associated pensions plan are thus applicable only to a very small proportion of the older adults employed in the public, Government, and Non-Government sectors of the economy [83], and in comparisons, such old age plans are more defined and well operated in a few Asian, American, and most of the European regions. In North Africa, only Algeria and Egypt have established national social pension schemes, and the latest available labour statistics suggest that virtually all older workers aged 65 years and over are engaged with informal income sources [84, 85].

Another major concern throughout Africa is the incidence of high mortality rates in any given age, and an ageing African is twice as likely to be burdened with diseases in comparison to the rest of the World. DALY (Disability-adjusted life years) rates are higher in older men than women, and Cardiovascular diseases (CVD) are the most frequent cause of death amongst older adults aged 60 years and above, accounting for over one-quarter of DALYs within the older age group [86]. Diabetes, cancers, parasites, and vector-borne diseases are other common conditions affecting older Africans to a greater degree [87]. Prostate cancer, is a major contributor towards mortality amongst older men in Africa, accounting for the largest share of deaths in comparison to other older men in the World [21]. Living conditions, including water and sanitation, also play a major role in increasing the frequency of diarrheal diseases and even deaths [88], thus a continuous drive to improve the living conditions will perhaps begin to decrease the burden of deaths due to diarrheal disease, a problem particularly prevalent in Africa, apart from musculoskeletal conditions, hearing and vision loss, and urinary incontinences as the other disabling contributors amongst the older adults.

1.7.2 Sub-Saharan African

Sub-Saharan African region has an estimated total of one billion population approximately, and 34 out of the 49 Sub-Saharan countries, are amongst the least developed countries (LDCs) in the world, with substantially poor healthcare facilities and general well-being index for the overall population [89]. Approximately, 50% of the population lives below the poverty line of $1.90 per day, which further decreased during the COVID-19 pandemic (Fig. 1.8). UN report estimated 36.6 million older adults above the age of 60 years live in the Sub-Saharan region, and geographically, the Sub-Sharan region is classified as Eastern, Middle, Southern, and Western Africa.

Eastern African countries comprises of Burundi, Comoros, Djibouti, Eritrea, Ethiopia, French Southern Territories, Kenya, Madagascar, Malawi, Mauritius, Mayotte, Mozambique, Reunion, Rwanda, Seychelles, Somalia, South Sudan, Uganda, United Republic of Tanzania, Zambia, and Zimbabwe. Amongst them, only Mauritius and Seychelles have functioning support and old age care policies and system [84], along with a few privately operated charities. Family support in the

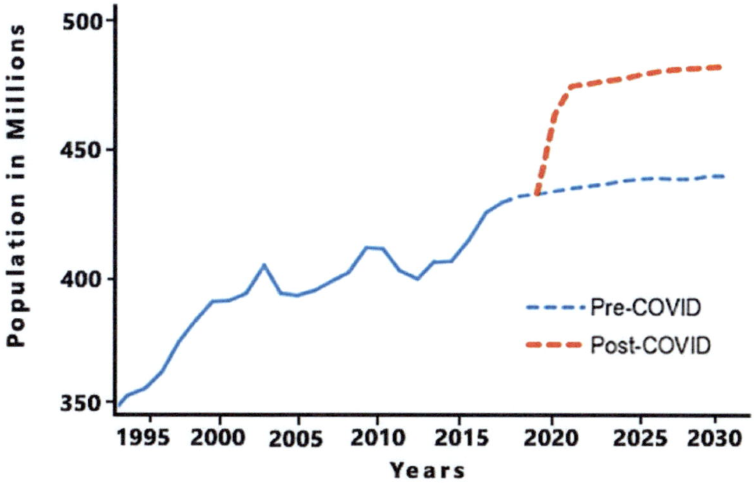

Fig. 1.8 COVID-19 pandemic caused more than 40 million people to enter into extreme poverty in Africa, living below $1.90 per day. Maharaj et al. 2020, and International Monetary Fund's calculations, 2021 [83, 95]

form of old age care for many of the Eastern Africa countries is a difficult prospect, as the cost associated with care is considered unaffordable for many communities and countries [90]. Thus, 96% of the older adults in East Africa remain as manual labour, and less than 22% of the are covered by pension plans or any other social welfare schemes [82]. Coverage for basic survival necessities such as food, water, and electricity is still inadequate in many rural areas of East Africa due to extreme poverty, thus proper healthcare facilities are out of reach, and focused care programmes for the elderly are still at an incipient stage. Lack of formal education is a persistent problem, and with higher mortality rates due to Human Immunodeficiency Virus (HIV) leading to Acquired Immunodeficiency Syndrome (AIDS) the age distribution pattern is thoroughly compromised, subsequently creating a deficit for care system offered through family [91]. Approximately 2.3 million people died of AIDS in 2004 [92], and by 2020, 75 million Africans have lost their lives to AIDS since the beginning of the epidemic [93]. Nevertheless, the number of older people is projected to nearly double in another 30 years.

Middle African countries comprises as Angola, Cameroon, Central African Republic, Chad, Congo, Democratic Republic of the Congo, Equatorial Guinea, Gabon, Sao Tome, and Principe. As demonstrated by Gross Domestic Product (GDP) calculations, the International Monetary Funds (IMF), and the World Bank data suggest that the standard of living in the Middle African countries tends to be the poorest amongst all the other World regions, and reliable data on income, expenditure, healthcare and general well-being for the older aged individuals is non-existent [94]. Middle African countries are still yet to develop standard pension plans or social welfare schemes for the aged, and the post-COVID World Bank report [95] shows majority of the Middle African household fall below the poverty line except Equatorial Guinea, Sao Tome and Principe with abundant natural oil and coca production resources, but certain households even in these countries are still poor due to the impact of HIV/AIDS epidemic, violent internal conflicts, and lack of basic medical care facilities.

Southern Africa comprises as Botswana, Lesotho, Namibia, South Africa, and Swaziland. South Africa and Botswana are the two countries, with an 'upper-middle economy', and are also the two most richest countries in the African continent. South Africa also has the highest number of older individuals aged 60 years and above [8]. Botswana, Lesotho, Namibia, and South Africa operate with established social pension and many other aged care schemes that offer comprehensive coverage [84]. As a broad developmental goal to end rural household poverty, the South African Government has introduced financial welfare schemes to older women and men who were never employed previously with any type of government or private sector jobs, thus this scheme specially funds as income source for families without any active earning members. Transfer of funds is also assisted for household with children and older aged individuals to protect them from healthcare and food costs. Lesotho, Namibia, and Swaziland are 'low-middle income' countries that mainly export natural resources and manufactured goods, and even though with progressive economic growth in these countries, established old age pension plans lacks including education outreach, optimal healthcare services, and high incidence of AIDS which disrupts the economy as manpower loss [10]. In Swaziland approximately 40-45% of its population is affected by AIDS, thus morality rates are high and average life expectancy range from 40 to 55 years.

Western Africa comprises of Benin, Burkina Faso, Cabo Verde, Cote d'Ivoire, Gambia, Ghana, Guinea, Guinea-Bissau, Liberia, Mali, Mauritania, Niger, Nigeria, Saint Helena, Senegal, Sierra Leone, and Togo. Ageing population report by several bodies including the United Nations, confirms that Cabo Verde is the only West African country with a national old age pension scheme, thus labour force participation rates in older ages are extremely high ranging from 70 to 90%. Much like the rest of the developing low-income countries many underprivileged West African countries, grapple in the same context of inadequate food supply, poverty, lack of formal education, adequate healthcare, and social welfare services.

Remarkably, of the existing old age pension schemes being operated in the African countries, only six countries qualify for the universal prospect and they are South Africa, Botswana, Kenya, Mauritius, Namibia, and Seychelles [96].

1.8 Ageing Trends: Europe

European continent reported the highest number of deaths during the COVID-19 pandemic in December 2020, declining the total population of the European countries to 446.1 million as reported by the Eurostat population bureau, revised UN World population prospect, 2022 [8], and the World Bank report, 2021 [95]. The sudden and drastic increase in the death rates may have far-ranging implications for

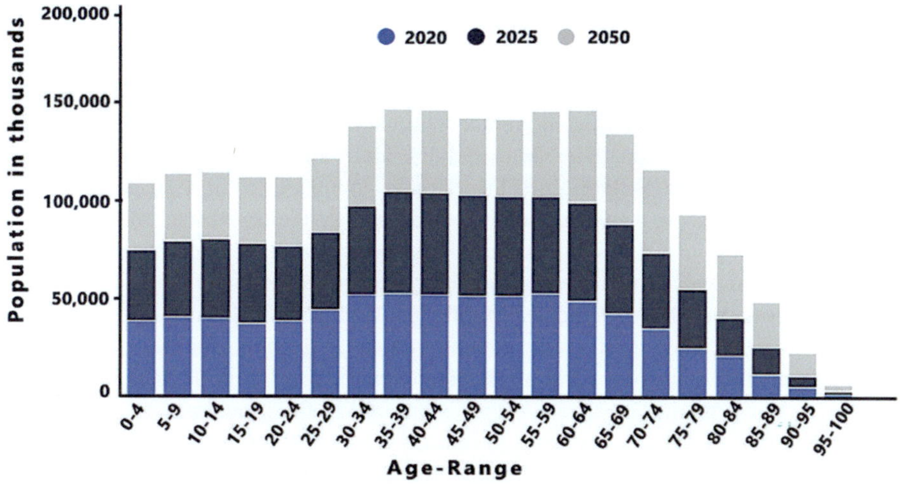

Fig. 1.9 Projected ageing population of the European continent, data extracted from European Ageing Network, 2020 [99]

the older society and the European economy, as recent studies on Corona Virus Disease (COVID-19) pandemic have widely reported higher fatality rates among older people than in the younger age groups with relevance to the infection-fatality ratio [97], and this fact also contributes to the European ageing pattern as perhaps gradually beginning to 'shrink prematurely' in its demographic regions. Approximately 20% of the older individual aged 65 years and above formed the total population of Europe before the COVID-19 pandemic, thus the European population are predicted to shrink gradually till 2050 (World population prospect, 2019, 2022).

The need for amendments to elderly care policies, therefore, increases as the European continent not only require income securities for the oldest old unable to work, but may also have to increase access towards affordable healthcare, social care, and long term, especially care programme [98] for elderlies above the age of 80 years (Fig. 1.9). Other factors, influencing the well-being of the aged apart from implementing standard country-wise pension policies are flexible re-employment opportunities, age-friendly housing, transportation, and easy access to public spaces [99]. 'Active ageing policies' is a recently coordinated cooperation between the public and private sectors, and the driving impetus behind this ageing policy in Europe is to ensure adequate provision of eldercare is fostered for both the 'younger old' and the frail 'oldest-old' (Fig. 1.9).

Ageing trends, polices, and access to socio-economic benefits like the rest of the World are not uniform in the European countries, and these differences can be explained and analysed by considering the various sub-divisions of the European continent categorised as Central, Eastern, Northern, Western, and Southern European countries or regions.

1.8.1 Central and Eastern European Countries

Central and Eastern European countries comprises of Germany, Poland, Czech Republic, Austria, Slovakia, Slovenia, Hungary, Belarus, Bulgaria, Moldova, Poland, Romania, Russia, Slovakia, and Ukraine, and amongst them Moldova, Ukraine, Belarus, and Bulgaria are the poorest countries of the European region, as these countries were a formal member of the Soviet Union Russia, thus developmental growth opportunities were markedly slower than other countries [100].

The standard ageing welfare regimes being practiced in many East and Central European countries have dedicated public pension policies, healthcare facilities, long-term care plans, and even educational growth opportunities leading to re-employment for the older community, however, exceptions are Bulgaria, Georgia, Moldova, and Romania, as these countries are still in the process of achieving full implementation of standard welfare policies, therefore, the prevailing old age care culture are similar to the Asian continent where the elders cohabit and reside in large multigenerational households, and as a cultural norm of respect [100, 101] the families, therefore care for the parents and elderly members. and they provide 'unpaid care services' as long as possible, as eldercare still suffers from lack of infrastructure investment, innovative practices in terms of healthcare facilities, and progressive social welfare policies [102].

1.8.2 Southern European

Southern European countries such as Cyprus, Greece, Portugal, and Spain, similarly rely on co-residential care as the foremost support towards the elderlies. In a few Southern European countries, informal home care duties are frequently outsourced by families to hire migrant carers through the 'grey market', this type of process or arrangments is unpaid by the Government and often delivered by inexperienced carers. According to the OECD, in low-service European countries, the expansion of informal old aged care services appears inevitable so as to relieve the burden on family and help them reconcile with work, in an attempt to meet the cost and increase the quality of care for the oldest old [103, 104].

Another important focus area for elderly care in Southern European countries has been to provide long-term care for the frail oldest-old, given the rapid increase in the number of the oldest old elderly in recent times. Further, individuals aged 80 years or above are projected to almost triple to 11% after 2060 [8], thus developing long-term care is an unavoidable issue and is requires well-coordinated orientation in Cyprus, Greece, Portugal, and Spain. Currently, these countries are also thriving with the family care culture reforms, especially in the rural regions. Thus, in this context, the need to reform long-term care provision, and to further formalise it has become one of the essential factors for aged care in Central, Eastern, and South European countries.

1.8.3 Northern and Western European Countries

Northern and Western European countries have the largest number of older aged retiring populations, and include economies from Denmark, Finland, Ireland, Norway, Sweden, and the United Kingdom as the Northern Europe region, and Austria, France, Germany, the Netherlands, and Switzerland are the Western European countries.

In all of these countries, a functional standard pension plan is already administered since 2002 [10], due to a relatively higher proportion of older working individuals who are still employed, and this has been attainable through advanced healthcare facilities, educational outreach, and smart technologies, still being implemented in most Asian and African continents. Fifty percent of the older people are employed in these countries since the 1990s [105], but, employment alone does not solve the need of care for the frail and the oldest old, as they require trained geriatric specialists to look after them, shortage in expert manpower for elder care is thus one of the main issues concerning ageing in Europe [106], and this is perhaps were countries such as China, India, and Africa are progressing towards in terms of ageing trends in the future, as widely reported by the UN world population prospect report series since the 1990s.

'Western and Northern European' countries are therefore live examples of how an ageing society and economy contributes towards progressively increasing the 'oldest-old age dependency ratio' that measures how many people should work to support those who are too old, or too young, to sustain the economy. The ratio is typically calculated by dividing the number of people who are under 16 years or over 65 years, by the number of people who are of 'working ages' between 16 and 64 years [107]. The old-age dependency ratio of Northern and Western Europe was already at 26.1 in 2010, which means there were roughly 4 working-age adults per dependent. But, by 2050, Europe's dependency ratio is projected to reach 50.2 which will be just 2 working-age adults per dependent [67].

Currently, the Netherlands and the Scandinavian countries spend 3–4 percent of their GDP on elder care, the highest in the World, apart from the United States, Australia, and Japan [108]. However, simply making funds available for long-term and aged care has been considered as in-effective approach towards improving the well-being of the oldest old, who are vulnerable to recurring illness, immobility, and are left 'alone', thus care homes facilities available to cater to their needs are often left empty in the rich Western and Northern European countries, as it was widely reported that treating the frail and weak with medical support is a last minute need, and providing a positive, caring, and supportive unit were essential prerequisite for the oldest old along with the administration of recommended medical treatment [109]. Social and mental healthcare support is also reported as a service that lacks trained manpower and sufficient care [110]. Western European countries are the first to address such needs, and to solve these issues the European Ageing Network (EAN) affiliated to the Global Ageing Network (GAN) has been formed [111, 112]. The 'vision and mission' of EAN and GAN is to develop innovative ideas for senior care with a practical approach, and some of them being developed include, advanced home automation and detection systems to increase the independence of older vulnerable who are living alone or are dependent on informal and professional care. Augmented reality, is being considered for further development to support the oldest-old dementia suffers, such advanced services are being built around the owned houses of elderlies by creating almost real-life assistances based on memories from their earlier life. Robotics is being explored to aid in companionship to older suffering from dementia and loneliness whilst offering care and monitoring. Tele-health services to rural and remote settings are already being practiced in full scale to provide care and assistance from a distance, which are some of the valuable live-changing support aids being explored to better serve the needs of the ageing community in the European continent.

1.9 Ageing Trends: Northern American Continent

The ‘Northern American’ continent was at 591 million population wise in 2022, and for the purpose of description the UN statistics division considers ‘Northern American’ region comprising of North America, Central America, and the Caribbean, with a total of 23 countries. It may also be noted that often a distinctive term ‘North America’ is used to demonstrate free trade agreement between the United States, Mexico, and Canada, which also forms the largest and the most populated ‘Northern American’ countries. Thus, the United States form 55% of the total population, Mexico is at 20% and Canada at 5%, followed by less than 4–0.4% are Guatemala, Haiti, Dominican Republic, Cuba, Honduras, Nicaragua, El Salvador, Costa Rica, Panama, Jamaica, Trinidad and Tobago, Bahamas, Belize, Barbados, Saint Lucia, Grenada, Saint Vincent and the Grenadines, Antigua, Barbuda, Dominica, Greenland and Saint Kitts, and the Nevis [8].

In Northern America, the United States alone make-up at least 20% of the total population as the ‘aged and retired’ residing in one-quarter of the US counties [113]. The United States, is therefore, the second largest country in the World to foster the highest number of older ageing population, after Japan, being the first. Older adults in North America are not only living longer but they are also more healthier than the previous generations. The increased longevity of life within a population is determined largely by analysing the declining death rates caused due to severe illness, or due to decreased physical and mental functions as indicated by the recent data trends compared to the 1960s (Fig. 1.10). However, increase in life expectancy is often accompanied by chronic disabilities and lack of adequate social well-being care for the aged. Much like the European continent, the standard of implementation of aged care policies is non-uniform across the Northern American countries. Although Social Security Scheme including pensions keeps most of the older Northern Americans out of poverty, there are wide economic disparities between different countries and the population such as the non-white Hispanic American, Latin American, and African American ethnicities.

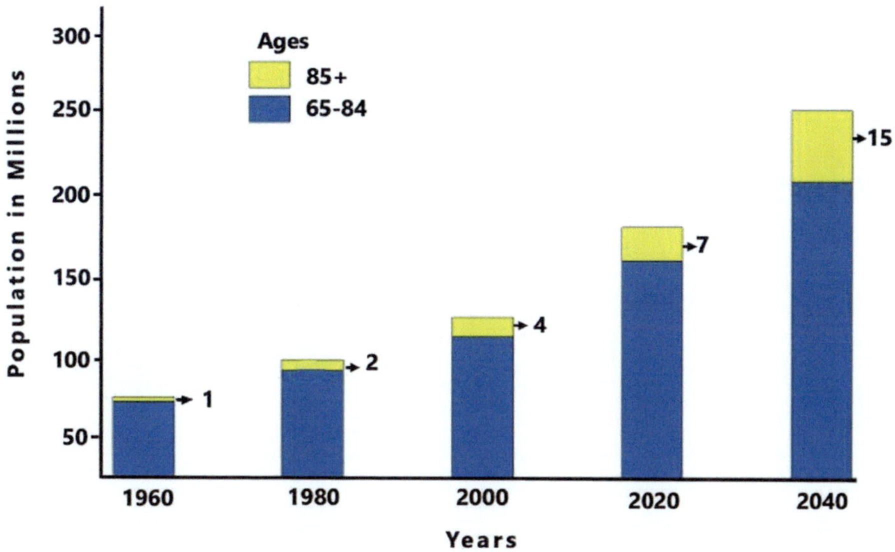

Fig. 1.10 Increasing ageing population and the oldest old group amongst the older Americans, 1960–2040, UN population database [71]

Older Northern American women are especially more vulnerable to differences in ageing care reforms, and in 2014, about 12% of the women aged 65 years and older were poor, compared to 7% of older men [114]. Amongst ages 75 years and older, women were nearly twice as likely to be poor (15%) compared to men (8%). Such differences at older ages can be explained, in part, by longer life expectancy experienced by women in comparison to men [115], and the fact that older women are more likely to live alone [116], and higher poverty amongst older women has also been linked to the gender gap in earnings at younger ages, which translates as lower social security earnings for older women during retirement [117].

Apart from the planned ageing policies implemented in North America (Fig. 1.10), many of the other Northern American countries are still not well established with aged care facilities, and social welfare scheme including pension plan which is received by only 4–10% of the older adults. The main reason behind the lack of pension support is largely due to the lack of proper Government and private sectors-endorsed employment schemes, as many older adults, especially the migrated sector, rely on informal job market, agricultural work, and manual labour during their lifetime to meet their financial needs. Thus, a voluntary pension scheme is available only to those older adults who had a properly endorsed job before the retirement age [118] (Fig. 1.11).

In El Salvador, as an example, 16% of the older adults report receiving a social pension, 22% of them are older men in comparison to women who are only at 7%. Twenty-one percent of older persons with disabilities receive pensions, and 5% of them receive non-contributory pensions as war veterans [119]. In Haiti and

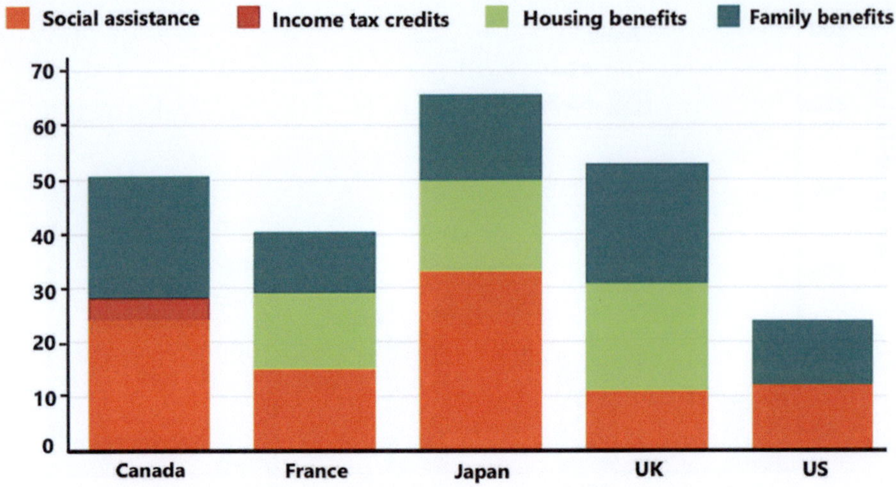

Fig. 1.11 Comparisons of welfare benefits received by the ageing group in high-income countries apart from pensions. Welfare, in the United States is categorised as family benefit by the OECD, as it is only available to those taking care of children. Food stamps are categorised as social assistance. OCED Report [9]

Honduras, 91% of the older adults confirm receiving no pension at all, like many other countries in Asia, Europe, and Sub-Saharan Africa [96].

Thus, the 'social hallmarks of aging' in Northern American countries apart from the United States, Canada, Bahamas, Mexico, Costa Rica, Saint Lucia, Antigua, and Puerto Rico are low lifetime socio-economic status, poverty, and extreme food shortages, especially in Haiti and Venezuela. In Ecuador, Colombia, Peru, and El Salvador, the budget allocated to social pensions averages only 0.16% of GDP, which is in sharp contrast to 3–4% of GDP being spent on older adults in the European region. These facts clearly demonstrate that even though the average life expectancy is increasing across the World and in the Northern American continent, the quality of life is still poor with a lack of adequate healthcare facilities, and long-term care programmes for the oldest old [5, 119, 120]. Haiti, Nicaragua, and Honduras are some of the poorest countries in North-Central America, and the World, experiencing rapid population growth, which outpaces the country's ability to provide for its current population. Some of these countries suffer today owing to their long history of political instability, internal revolution, deforestation, natural disaster, corruption, debt, and violence contributing to poverty, and lack of essential infrastructure, thus many residents are without electricity, safe water, roads, and educational options. Many of the Central and South American countries are dependent on tourism for their economic prosperity and they suffer today due to COVID-19 pandemic, as a negative impact was set to disrupt the lives of many small businesses increasing unemployment after 2020. The World Bank data report suggests that the countries primarily based on tourism will sustain losses until

2025 [95]. Health and Healthcare facilities for all ages are therefore constrained. Many aged care facilities are available based on personal earnings, and economic standards of the business. Therefore, eventhough the average life expectancy in the Northern America countries has increased as a whole, but it is accompanied by increased disability, unavailability of adequate social reforms measures, thereby increasing dependence of the elderly care on other members of their families.

1.10 Ageing Trends: Latin America

Latin America is defined as all of the Americas south of the United States, with geographical subregions distinguished as North, Central, and South America (and the Caribbean). It may be noted that considerable overlapping of countries exists between the Northern American and Latin American continents, as there is no universally accepted division between the Latin American and the Northern American continent, therefore Latin America comprises of fifty countries in total [121].

North America and Caribbean countries of Latin America are Anguilla, Antigua and Barbuda, Aruba, Bahamas, Barbados, Bonaire, Sint Eustatius and Saba, British Virgin Islands, Cayman Islands, Cuba, Curaçao, Dominica, Dominican Republic, Grenada, Guadeloupe, Haiti, Jamaica, Martinique, Montserrat, Puerto Rico, Saint Barthelemy, Saint Kitts and Nevis, Saint Lucia, Saint Martin, Saint Vincent and the Grenadines, Sint Maarten, Trinidad and Tobago, Turks and Caicos Islands, and United States Virgin Islands.

Central American countries are Belize, Costa Rica, El Salvador, Guatemala, Honduras, Mexico, Nicaragua, and Panama.

South American countries of Latin America comprises Argentina, Bolivarian Republic of Venezuela, Brazil, Chile, Colombia, Ecuador, the Falkland Islands (Malvinas), French Guiana, Guyana, Paraguay, Peru, Plurinational State of Bolivia, Suriname, and Uruguay.

In 2022, the total population of Latin America continent was reported to be 660 million, and 58.6 million people aged 60 years and above, represented 13.4% of the retired and aged population [8]. Furthermore, women formed a larger proportion of aged individuals than men due to longer life expectancy experienced by the females, similar to Northern American continent. Further by 2025, Latin American countries are projected to progress as a young adult society, before becoming an adult society in 2045, with highest population in the age range of 40–59 years [122]. Thus, in less than 10 years, older population aged 60 years and above, will therefore begin to supersede all other age groups. These estimates suggest that considerable restructuring of the aged public policy is an impending need (Fig. 1.12) and considering the current ageing-related constraints faced by the United States, Japan, and European countries, in the foreseeable future Latin America may have to introduce many new long-term plans for pension polices including labour benefits, health, social welfare, and other elderly care policies.

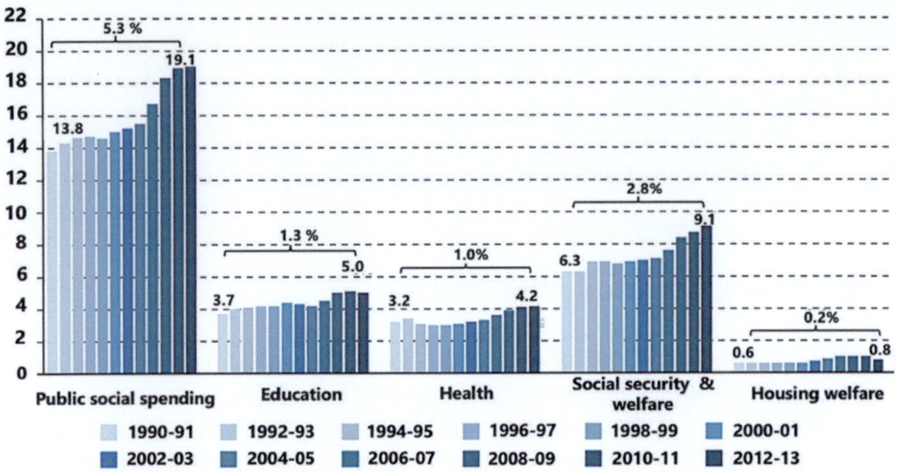

Fig. 1.12 Latin America and Caribbean countries: sector-wise social spending for public reform suggesting relative need for improvements in health and housing welfare, reproduced from Economic commission for Latin America and Caribbean countries [123]

In general, ageing process and implementation of current age-related welfare policies is not homogeneous in many countries and continent in the World, and similarly in Latin America differences exists between countries, including urban and rural areas, between cities, and even between population sub-groups, such as the indigenous inhabitants, persons of migratory descent, who are often subjected to certain levels of social and economic inequality [123].

Contrasting with the Latin American and the Caribbean region, the South American country of Uruguay exceeds 20% of its population as 'aged and retried' [124], and Chile is expected to reach that mark by 2030, which will make them the oldest populations within the subregion. Other South American countries following similar ageing trends are Argentina, Brazil, and Colombia. Amongst the Central American countries, Costa Rica is undergoing the most rapid population ageing process increasing from the current 16–20% in this decade [8]. Countries lagging the most in terms of ageing population and social security policies are Belize, Guatemala, Honduras, and Nicaragua, which are also the most economically deprived Latin American countries with adverse environmental, political, educational, and infrastructural conditions. Further, COVID-19 pandemic has additionally contributed towards descaling the economic growth of the many Central American countries that were heavily dependent on tourism as a source of employment and earnings.

Therefore, the fundamental gap that still requires considerable attention to let the ageing community survive successfully in Latin America involves, improving the present and past employment trajectories, reducing inequalities in female wages, implementing adequate housing facilities towards migrant older persons, the nationwide launch of social protection schemes for the aged residing in rural areas, adequate support towards social well-being, and improved healthcare facilities and opportunities for all races and ethnicities.

1.11 Ageing Trends: Oceanic Continent

The total population of the Oceanic continent in 2022 was reported as 44 million approximately, and 13.4% of them formed the 'aged and retired' [125]. For the purpose of description, the Oceanic continent is grouped into four subregions named as Australasia comprising the countries Australia and New Zealand, Melanesia comprises Papua New Guinea, Fiji, Solomon Islands, Vanuatu, and New Caledonia, Micronesian countries comprise of Federated States of Micronesia and Guam, and Polynesia includes French Polynesia, Samoa, Tonga, Tokalau, and Niue [53, 126].

Much like the rest of the World, the Oceanic continent presents a mix of countries that are either ranked highly developed in life quality and human development index, as demonstrated by Australia (Fig. 1.13), French Polynesia, Hawaii, New Caledonia, and New Zealand, to the much less developed regions in an urgent need to develop heath and community care centres at Papua New Guinea, Tuvalu, Vanuatu, Western New Guinea, including moderate economics, such as Fiji, Palau, and Tonga [70, 127]. Similar, to many other countries in Europe, Latin America, and Northern America, women tend to have a longer life span and lower death rates in comparison to men in the Oceanic countries, and similar to many Asian and African countries, the primary care for the older individuals is offered through family home as a strongly held value for elderlies in low and middle income Oceanic countries.

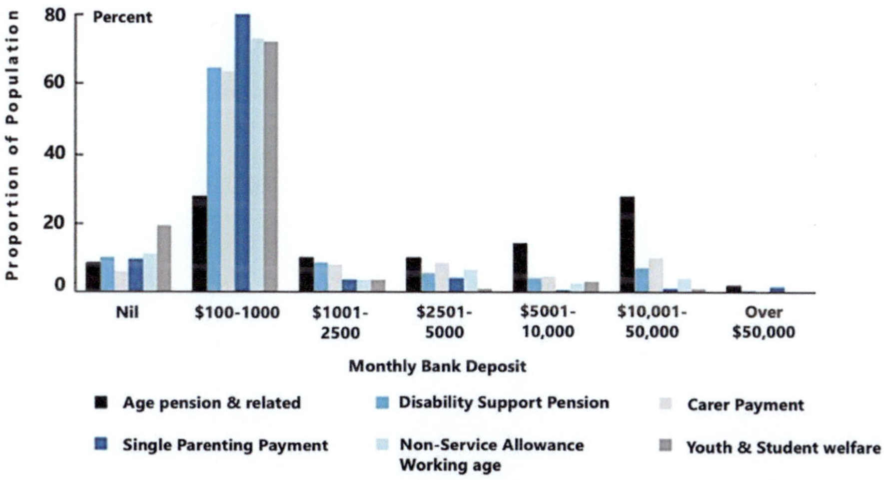

Fig. 1.13 Highly reliant income support recipient in Australia, the best in human development index. Projected data source as verified by OECD, and the Australian Institute of Health and Welfare [124]

However, as life expectancy increases globally unpaid family-driven informal home care support is becoming a less feasible approach, as the oldest old with disabilities often require constant attention by a trained geriatric expert. Such assistance is not available in the low and middle-income Oceanic countries, due to the accompanying economic pressure on the Government. Moreover, informal caregiving, particularly for the oldest old, is stressful for unpaid caregivers who can easily become overburdened, especially when attempting to balance child care and employment obligations.

Social welfare scheme includes pension plans which range from being contributory, non-contributory, or may include voluntary coverage to cover a certain period of employment. Pensions are received by 30% of the older inhabitants in the Oceanic countries which indicates that continent-wide comprehensive coverage does not exist. Papua New Guinea and Fiji have a National provident fund scheme to which both public and private employees contribute. Australia, New Zealand, and Hawaii island have the best old age pension policies [8, 128, 129], and Non-contributory old age pensions benefits are paid in Niue and Samoa, but the amounts are small due to financial constraints, and attempts to increase the old age pension has not been a positive resolution as in the past, natural disasters such as cyclones, tsunamis, droughts, floods, and earthquakes were common amongst the oceanic continent. Rural village workers and urban informal sector workers are largely uncovered by contributory schemes but may have access to small sum of financial government assistance.

Health services required for an ageing population tend to be both labour and technology intensive, and in developing Oceanic countries health services are inadequate and overburdened. In moderate and low-income Oceanic countries, the resurgence of infectious communicable diseases, such as the COVID-19 pandemic occurred when other non-communicable diseases were also on rise, thereby creating a double burden for health services (Fig. 1.14). Most countries have established referral programmes whereby patients are sent to more advanced medical centres in Australia, Hawaii, or New Zealand for treatments that are not unavailable in their home country, but due to high costs associated with referral to overseas facilities access is highly restricted, and elderly patients are unlikely to receive priority through such schemes which are generally focused on children and working-age persons, although there is no discrimination on the basis of age alone.

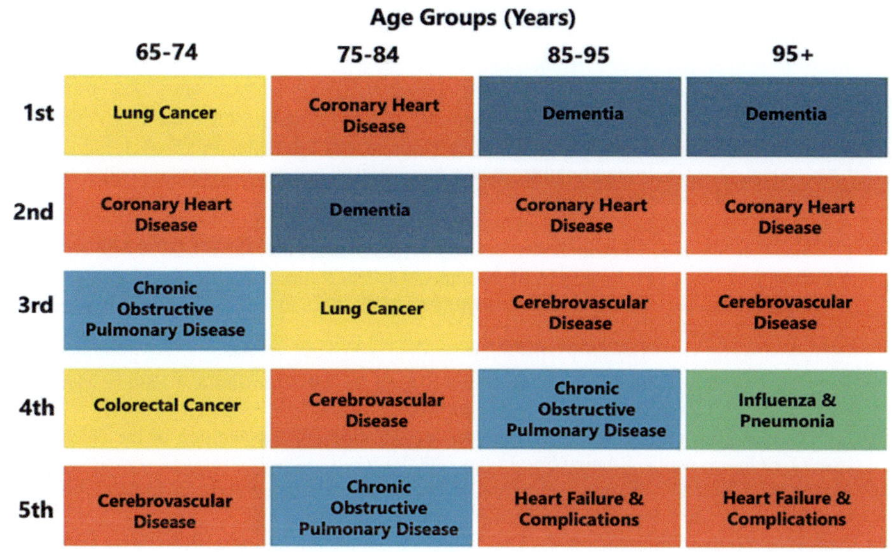

Fig. 1.14 Five leading causes of death in older adults categorised through age groups in the Oceanic countries, as determined by the Australian Institute of Health and Welfare [125]

Competing Interest The author has no conflicts of interest to declare that are relevant to the content of this chapter.

References

1. Sanderson WC, Scherbov S. Demography. Remeasuring aging. Science. 2010;329(5997): 1287–8.
2. Organisation for Economic Co-operation and D. Pensions at a glance 2021. S.l: OECD Publishing; 2021.
3. Nagarajan NR, Sixsmith A. Policy initiatives to address the challenges of an older population in the workforce. Ageing Int. 2023;48(1):41–77.
4. The retirement effects of old-age pension and early retirement schemes in OECD countries. OECD Economics Department Working Papers. ISSN: 18151973 (online). https://doi.org/10.1787/18151973
5. Brown GC. Living too long: the current focus of medical research on increasing the quantity, rather than the quality, of life is damaging our health and harming the economy. EMBO Rep. 2015;16(2):137–41.
6. Lee R, Mason A. Some macroeconomic aspects of global population aging. Demography. 2010;47(Suppl 1):S151-72.
7. Hellevik S, Mustafa S, Zhang Y, Shirsat A, Saikat S. Multisectoral action towards sustainable development goal 3.d and building health systems resilience during and beyond COVID-19: findings from an INTOSAI development initiative and World Health Organization collaboration. Front Public Health. 2023;11:1104669.
8. United N. World population prospects 2022 : summary of results. New York: United Nations; 2022.

9. Organisation for Economic Co-operation and D. Working better with age. Paris, France: OECD; 2019.
10. Jivetti B, Hoque MN. SpringerLink. Population change and public policy. 1st 2020. Ed. Cham: Springer International Publishing : Imprint: Springer; 2020.
11. Rajan SI, Sarma PS, Mishra US. Demography of Indian aging, 2001–2051. J Aging Soc Policy. 2003;15(2–3):11–30.
12. Chadda RK, Deb KS. Indian family systems, collectivistic society and psychotherapy. Indian J Psychiatry. 2013;55(Suppl 2):S299-309.
13. Asaria M, Mazumdar S, Chowdhury S, Mazumdar P, Mukhopadhyay A, Gupta I. Socioeconomic inequality in life expectancy in India. BMJ Glob Health. 2019;4(3):e001445.
14. Vaishnav LM, Joshi SH, Joshi AU, Mehendale AM. The National Programme for Health Care of the Elderly: a review of its achievements and challenges in India. Ann Geriatr Med Res. 2022;26(3):183–95.
15. Dumka N, Mangat S, Ahmed T, Hannah E, Kotwal A. Adding health to years: a review of the National Programme for Health Care of the Elderly (NPHCE) in India. J Family Med Prim Care. 2022;11(11):6654–9.
16. Sawhney M. The role of non-governmental organizations for the welfare of the elderly: the case of HelpAge India. J Aging Soc Policy. 2003;15(2–3):179–91.
17. Verma R, Khanna P. National program of health-care for the elderly in India: a hope for healthy ageing. Int J Prev Med. 2013;4(10):1103–7.
18. East-West Center Program on P. The future of population in Asia. Honolulu: East-West Center; 2002. p. 150. col.ill., graphs,maps,tables p
19. Salehi M, Aziz-Mohammdlooha M, Masaebi F, Zayeri F. Six-decade trend analysis of life expectancy at birth in Asia, regions of Asia and Iran: a joinpoint analysis. Med J Islam Repub Iran. 2022;36:166.
20. Roy S, Khatun T. Effect of adolescent female fertility and healthcare spending on maternal and neonatal mortality in low resource setting of South Asia. Health Econ Rev. 2022;12(1):47.
21. Collaborators GCoD. Global, regional, and national age-sex-specific mortality for 282 causes of death in 195 countries and territories, 1980-2017: a systematic analysis for the Global Burden of Disease Study 2017. Lancet. 2018;392(10159):1736–88.
22. Knaus JP. The problem of overpopulation in Asia. Ther Ggw. 1981;120(6):499–516.
23. Tan YL, Chen JE, Yiew TH, Habibullah MS. Habitat change and biodiversity loss in South and Southeast Asian countries. Environ Sci Pollut Res Int. 2022;29(42):63260–76.
24. Kruk ME, Gage AD, Joseph NT, Danaei G, García-Saisó S, Salomon JA. Mortality due to low-quality health systems in the universal health coverage era: a systematic analysis of amenable deaths in 137 countries. Lancet. 2018;392(10160):2203–12.
25. Renwick N. Millennium Development Goal 1: poverty, hunger and decent work in Southeast Asia. Third World Q. 2011;32(1):65–89.
26. Chan A. Aging in Southeast and East Asia: issues and policy directions. J Cross Cult Gerontol. 2005;20(4):269–84.
27. Fang Y, Zhang F. The future path to China's poverty reduction-dynamic decomposition analysis with the evolution of China's poverty reduction policies. Soc Indic Res. 2021;158 (2):507–38.
28. Ye M, Zhang J, Li H. Twins, income, and happiness: evidence from China. Proc Natl Acad Sci U S A. 2023;120(25):e2221884120.
29. Zheng X, Luo Y, Su B, He P, Guo C, Tian Y, et al. Developmental gerontology and active population aging in China. China CDC Wkly. 2023;5(8):184–7.
30. Wang H, Chen H. Aging in China: challenges and opportunities. China CDC Wkly. 2022;4 (27):601–2.
31. Li J, Lai DWL, Chappell NL. The older, the happier?: a cross-sectional study on the "paradox of aging" in China. Res Aging. 2022;44(1):34–43.
32. Liu L, Wu F, Tong H, Hao C, Xie T. The digital divide and active aging in China. Int J Environ Res Public Health. 2021;18:23.

33. Feng Z, Wu B. Embracing challenges for population aging in China: building scientific evidence to inform long-term care policymaking and practice. J Aging Soc Policy. 2023:1–11.
34. Xiao Y, Li J. A conversational analysis of aging in China from a cross-section of the labour market: a corpus-based study. Humanit Soc Sci Commun. 2022;9(1):377.
35. United N. Addressing the challenges of population ageing in Asia and the Pacific : implementation of the Madrid international plan of action on ageing. New York: United Nations; 2017.
36. Raymo JM, Park H, Xie Y, Yeung WJ. Marriage and family in East Asia: continuity and change. Annu Rev Sociol. 2015;41:471–92.
37. Yeung WJ, Lee Y. Aging in East Asia: new findings on retirement, health, and Well-being. J Gerontol B Psychol Sci Soc Sci. 2022;77(3):589–91.
38. Saito T, Izawa KP. Effectiveness and feasibility of home-based telerehabilitation for community-dwelling elderly people in Southeast Asian countries and regions: a systematic review. Aging Clin Exp Res. 2021;33(10):2657–69.
39. Ikegami N, Yoo BK, Hashimoto H, Matsumoto M, Ogata H, Babazono A, et al. Japanese universal health coverage: evolution, achievements, and challenges. Lancet. 2011;378(9796): 1106–15.
40. Arai H, Ouchi Y, Toba K, Endo T, Shimokado K, Tsubota K, et al. Japan as the front-runner of super-aged societies: perspectives from medicine and medical care in Japan. Geriatr Gerontol Int. 2015;15(6):673–87.
41. Christensen K, Doblhammer G, Rau R, Vaupel JW. Ageing populations: the challenges ahead. Lancet. 2009;374(9696):1196–208.
42. Population dynamics in Asia and the Pacific: implications for development. Asia Pac Popul J. 1994;9(1):57–75.
43. Smith JP, Majmundar MK, National Research Council (U.S.). Panel on Policy Research and Data Needs to Meet the Challenge of Aging in Asia. Aging in Asia : findings from new and emerging data initiatives. Washington, D.C.: National Academies Press; 2012. Available from: https://www.ncbi.nlm.nih.gov/books/NBK92618/
44. Eggleston K, Walter H. Shorenstein Asia-Pacific Research Center. Healthy aging in Asia. Stanford, CA: The Walter H. Shorenstein Asia-Pacific Research Center, Freeman Spogli Institute for International Studies, Stanford University; 2020. xxvi, 278 pages p
45. Sadigov R. Rapid growth of the world population and its socioeconomic results. ScientificWorldJournal. 2022;2022:8110229.
46. Kudrna G, Le T, Piggott J. Macro-demographics and ageing in emerging Asia: the case of Indonesia. J Popul Ageing. 2022;15(1):7–38.
47. Appelt JL, Garcia Rojas DC, Verburg PH, van Vliet J. Socioeconomic outcomes of agricultural land use change in Southeast Asia. Ambio. 2022;51(5):1094–109.
48. Sharma D. Achieving sustainable development nutrition targets: the challenge for South Asia. J Glob Health. 2020;10(1):010303.
49. Wagstaff A. Poverty and health sector inequalities. Bull World Health Organ. 2002;80(2): 97–105.
50. Muoneke OB, Okere KI, Nwaeze CN. Agriculture, globalization, and ecological footprint: the role of agriculture beyond the tipping point in The Philippines. Environ Sci Pollut Res Int. 2022;29(36):54652–76.
51. Zhang J, Zhao W, Cheng B, Li A, Wang Y, Yang N, et al. The impact of digital economy on the economic growth and the development strategies in the post-COVID-19 era: evidence from countries along the "belt and road". Front Public Health. 2022;10:856142.
52. Park D, ProQuest. Pension systems and old-age income support in East and Southeast Asia : overview and reform directions. Abingdon, Oxon ; New York: Routledge; 2011.
53. Devasahayam TW, Dhirathiti NS, Higo M, Klassen TR, Taylor F. Ageing in Asia-Pacific: interdisciplinary and comparative perspectives. 1st ed. New York: Routledge; 2018.
54. Sanchez-Romero M, Lee RD, Prskawetz A. Redistributive effects of different pension systems when longevity varies by socioeconomic status. J Econ Ageing. 2020:17.

55. Zhou Y, Zhang L, Xiao J, Williams CA, Vitkovskaya I, Bao A. Spatiotemporal transition of institutional and socioeconomic impacts on vegetation productivity in Central Asia over last three decades. Sci Total Environ. 2019;658:922–35.
56. Mathers CD, Stevens GA, Boerma T, White RA, Tobias MI. Causes of international increases in older age life expectancy. Lancet. 2015;385(9967):540–8.
57. Pulatova G, Harun-Or-Rashid M, Yoshida Y, Sakamoto J. Elderly health and its correlations among Uzbek population. Nagoya J Med Sci. 2012;74(1–2):71–82.
58. Adambekov S, Askarova S, Welburn SC, Goughnour SL, Konishi A, LaPorte R, et al. Publication productivity in Central Asia and countries of the Former Soviet Union. Cent Asian J Glob Health. 2016;5(1):261.
59. Wagstaff A, Moreno-Serra R. Social health insurance and labor market outcomes: evidence from central and eastern Europe, and Central Asia. Adv Health Econ Health Serv Res. 2009;21:83–106.
60. Mason KO. Family change and support of the elderly in Asia: what do we know? Asia Pac Popul J. 1992;7(3):13–32.
61. Goh VH. Aging in Asia: a cultural, socio-economical and historical perspective. Aging Male. 2005;8(2):90–6.
62. Takayama N, 高山 憲, 一橋大学経済研究所. Pensions in Asia : incentives, compliance and their role in retirement. Tokyo: Maruzen Co; 2005. xx, 256 pages : illustrations p.
63. West Asia. Nature. 2015;522(7556):S22-3.
64. Harpham T, Stephens C. Urbanization and health in developing countries. World Health Stat Q. 1991;44(2):62–9.
65. Skirbekk V, Dieleman JL, Stonawski M, Fejkiel K, Tyrovolas S, Chang AY. The health-adjusted dependency ratio as a new global measure of the burden of ageing: a population-based study. Lancet Healthy Longev. 2022;3(5):e332–e8.
66. Walker A, Foster L, Edward EP. The political economy of ageing and later life : critical perspectives. Cheltenham: Edward Elgar Pub. Ltd; 2014. 1 online resource (v.) p
67. World Population Prospects: United Nations. Department of Economic and Social Affairs. UN DESA/POP/2021/TR/NO. 3. ISBN: 978-92-1-148373-4
68. Ibrahim NA, Edis Z, Al-Owais KS. Adherence of geriatric patients and their beliefs toward their medicines in The United Arab Emirates. J Pharm Bioallied Sci. 2020;12(1):22–30.
69. Blair I, Sharif A. Health and health systems performance in The United Arab Emirates. World Hosp Health Serv. 2013;49(4):12–7.
70. Western M, Tomaszewski W. Subjective wellbeing, objective wellbeing and inequality in Australia. PLoS One. 2016;11(10):e0163345.
71. United Nations. Department of International Economic and Social A. World population prospects. Population studies. New York: United Nations.
72. Kalu ME, Ojembe BU, Akinrolie O, Okoh AC, Adandom II, Nwankwo HC, et al. Setting priorities for ageing research in Africa: a systematic mapping review of 512 studies from sub-Saharan Africa. J Glob Health. 2021;11:15002.
73. Masquelier B, Reniers G, Pison G. Divergences in trends in child and adult mortality in sub-Saharan Africa: survey evidence on the survival of children and siblings. Popul Stud (Camb). 2014;68(2):161–77.
74. Pourreza A, Sadeghi A, Amini-Rarani M, Khodayari-Zarnaq R, Jafari H. Contributing factors to the total fertility rate declining trend in the Middle East and North Africa: a systemic review. J Health Popul Nutr. 2021;40(1):11.
75. Alesina A, Hohmann S, Michalopoulos S, Papaioannou E. Religion and educational mobility in Africa. Nature. 2023;618(7963):134–43.
76. Jecker NS. Intergenerational ethics in Africa: duties to older adults in skipped generation households. Dev World Bioeth. 2022;22(3):152–61.
77. Kebede E, Goujon A, Lutz W. Stalls in Africa's fertility decline partly result from disruptions in female education. Proc Natl Acad Sci U S A. 2019;116(8):2891–6.

78. Moultrie TA, Sayi TS, Timæus IM. Birth intervals, postponement, and fertility decline in Africa: a new type of transition? Popul Stud (Camb). 2012;66(3):241–58.
79. Karlin NJ, Salem MB, Weil J. Aging in Tunisia. Gerontologist. 2018;58(6):1004–8.
80. Ben AA. 40 years of basic health care in Tunisia Alma Ata in Astana. It's time to revitalize the first line of care. Tunis Med. 2021;99(1):179–88.
81. Smith-Greenaway E. Educational attainment and adult literacy: a descriptive account of 31 sub-Saharan Africa countries. Demogr Res. 2015;33:1015–34.
82. Darkwa K, Mazibuko FN, Candidate PH. Population aging and its impact on elderly welfare in Africa. Int J Aging Hum Dev. 2002;54(2):107–23.
83. Maharaj P. Health and care in old age in Africa. London ; New York: Routledge; 2020. xiii, 160 pages p
84. Bonoli G, Shinkawa T. Ageing and pension reform around the world : evidence from eleven countries. Cheltenham, UK ; Northampton, MA: Edward Elgar; 2005. xi, 279 p. p
85. Aksoy MA. African agricultural reforms : the role of consensus and institutions. Washington, D.C.: World Bank; 2012. xix, 313 p. p
86. Collaborators GDaH. Global, regional, and national disability-adjusted life-years (DALYs) for 359 diseases and injuries and healthy life expectancy (HALE) for 195 countries and territories, 1990-2017: a systematic analysis for the Global Burden of Disease Study 2017. Lancet. 2018;392(10159):1859–922.
87. Setel PW. Non-communicable diseases, political economy, and culture in Africa: anthropological applications in an emerging pandemic. Ethn Dis. 2003;13(2 Suppl 2):S149-57.
88. Mutono N, Wright J, Mutembei H, Muema J, Thomas M, Mutunga M, et al. The nexus between improved water supply and water-borne diseases in urban areas in Africa: a scoping review protocol. AAS Open Res. 2020;3:12.
89. Adeyeye SAO, Ashaolu TJ, Bolaji OT, Abegunde TA, Omoyajowo AO. Africa and the nexus of poverty, malnutrition and diseases. Crit Rev Food Sci Nutr. 2023;63(5):641–56.
90. Kibirige JS. Population growth, poverty and health. Soc Sci Med. 1997;45(2):247–59.
91. Brennan-Ing M, Porter KE, Kaufman JE, MacPhail C, Seeley J. Aging with HIV in sub-Saharan Africa : health and psychosocial perspectives. Cham, Switzerland: Springer; 2022. xii, 223 pages p
92. National Research Council (U.S.). Panel on policy and data needs to meet the challenge of aging in Africa. In: Cohen B, Menken J, editors. Aging in sub-Saharan Africa : recommendations for furthering research. Washington, D.C.: National Academies Press; 2006. xii, 356 p. p.
93. Jamison DT, Feachem RG. Disease and mortality in Sub-Saharan Africa. 2nd ed. Washington, D.C.: World Bank; 2006. xxii, 387 p. p
94. Collier P. The bottom billion: why the poorest countries are failing and what can be done about it. Oxford ; New York: Oxford University Press; 2007. xiii, 205 p. p
95. World Bank (Washington District of Columbia). World Development Report 2021: data for better lives. Washington: World Bank; 2021. pages cm p
96. Publishing O. Organisation for economic co-operation and D. pensions at a glance : Asia/Pacific 2011. Paris: OECD; 2012.
97. Zhang H, Wu Y, He Y, Liu X, Liu M, Tang Y, et al. Age-related risk factors and complications of patients with COVID-19: a population-based retrospective study. Front Med (Lausanne). 2021;8:757459.
98. Calderón-Jaramillo M, Zueras P. Cared and uncared populations: understanding unmet care needs of older adults (65+) across different social care systems in Europe. Eur J Ageing. 2023;20(1):11.
99. Sánchez-Moreno E, Gallardo-Peralta LP. Income inequalities, social support and depressive symptoms among older adults in Europe: a multilevel cross-sectional study. Eur J Ageing. 2022;19(3):663–75.
100. Sompolska-Rzechuła A, Kurdyś-Kujawska A. Assessment of the development of poverty in EU countries. Int J Environ Res Public Health. 2022;19:7.

101. Weziak-Bialowolska D. Spatial variation in EU poverty with respect to health, education and living standards. Soc Indic Res. 2016;125(2):451–79.
102. Abbing HR. Health, healthcare and ageing populations in Europe, a human rights challenge for European health systems. Eur J Health Law. 2016;23(5):435–52.
103. Verbakel E. How to understand informal caregiving patterns in Europe? The role of formal long-term care provisions and family care norms. Scand J Public Health. 2018;46(4):436–47.
104. Bosch-Farré C, Garre-Olmo J, Bonmatí-Tomàs A, Malagón-Aguilera MC, Gelabert-Vilella S, Fuentes-Pumarola C, et al. Prevalence and related factors of active and healthy ageing in Europe according to two models: results from the survey of health, ageing and retirement in Europe (SHARE). PLoS One. 2018;13(10):e0206353.
105. Weber D, Loichinger E. Live longer, retire later? Developments of healthy life expectancies and working life expectancies between age 50–59 and age 60–69 in Europe. Eur J Ageing. 2022;19(1):75–93.
106. Stuck AE, Masud T. Health care for older adults in Europe: how has it evolved and what are the challenges? Age Ageing. 2022;51:12.
107. Muszyńska MM, Rau R. The old-age healthy dependency ratio in Europe. J Popul Ageing. 2012;5(3):151–62.
108. Papanicolas I, Woskie LR, Jha AK. Health care spending in the United States and other high-income countries. JAMA. 2018;319(10):1024–39.
109. Hamilton-West K, Milne A, Hotham S. New horizons in supporting older people's health and wellbeing: is social prescribing a way forward? Age Ageing. 2020;49(3):319–26.
110. Livingston G, Huntley J, Sommerlad A, Ames D, Ballard C, Banerjee S, et al. Dementia prevention, intervention, and care: 2020 report of the lancet commission. Lancet. 2020;396 (10248):413–46.
111. Green G. Age-friendly cities of Europe. J Urban Health. 2013;90(Suppl 1):116–28.
112. Van Hoof J, Kazak JK, Perek-Białas JM, Peek STM. The challenges of urban ageing: making cities age-friendly in Europe. Int J Environ Res Public Health. 2018;15:11.
113. Crimmins EM. Recent trends and increasing differences in life expectancy present opportunities for multidisciplinary research on aging. Nat Aging. 2021;1(1):12–3.
114. Carmel S. Health and well-being in late life: gender differences worldwide. Front Med (Lausanne). 2019;6:218.
115. Zarulli V, Barthold Jones JA, Oksuzyan A, Lindahl-Jacobsen R, Christensen K, Vaupel JW. Women live longer than men even during severe famines and epidemics. Proc Natl Acad Sci U S A. 2018;115(4):E832–E40.
116. Joe A, Dickins M, Enticott J, Ogrin R, Lowthian J. Community-dwelling older women: the association between living alone and use of a home nursing service. J Am Med Dir Assoc. 2020;21(9):1273–81.e2.
117. Mandel H, Semyonov M. Gender pay gap and employment sector: sources of earnings disparities in the United States, 1970-2010. Demography. 2014;51(5):1597–618.
118. Munnell AH, Hou W, College B. Center for Retirement Research. Will millennials be ready for retirement? Chestnut Hill, MA: Center for Retirement Research at Boston College; 2018. Available from: http://crr.bc.edu/wp-content/uploads/2018/01/IB_18-2.pdf, Available from: http://resource.nlm.nih.gov/101727768
119. Barrientos A. Inequalities in income security in later age in Latin America. Rev Panam Salud Publica. 2021;45:e85.
120. Crimmins EM. Lifespan and healthspan: past, present, and promise. Gerontologist. 2015;55 (6):901–11.
121. Wiarda JH, Siqueira WI. Latin America: population and internal unrest. Draper Fund Rep. 1985;14:20–4.
122. Castillo-Riquelme M, Yamada G, Diez Roux AV, Alfaro T, Flores-Alvarado S, Barrientos T, et al. Aging and self-reported health in 114 Latin American cities: gender and socio-economic inequalities. BMC Public Health. 2022;22(1):1499.

123. Keating NC, Rodríguez Mañas L, De Francisco A. Toward healthy aging in Latin America and the Caribbean: leaving no one behind? Rev Panam Salud Publica. 2021;45:e113.
124. Marroig A, Machado M, Muniz-Terrera G. Dependency change with aging and associated factors in Uruguay: a cohort study. J Aging Health. 2021;33(10):886–95.
125. CCH Australia Limited. Australian master superannuation guide. Sydney, NSW: CCH Australia. p. volumes
126. Arabena K, Kingsley J. Oceania. EcoHealth. 2016;13(4):615–7.
127. Koohi F, Nedjat S, Yaseri M, Cheraghi Z. Quality of life among general populations of different countries in the past 10 years, with a focus on human development index: a systematic review and meta-analysis. Iran J Public Health. 2017;46(1):12–22.
128. CCH Australia Limited. Australian superannuation legislation. North Ryde, NSW: CCH Australia. p. v.
129. Australia. Bureau of Census and Statistics. Survey of selected private pension funds.

Chapter 2
Leading Causes of Death and Disability Among the Global Aging Community

Bidita Khandelwal and Chamma Gupta

Abstract As longevity increases, diseases and disabilities will continue to rise progressively within the aging community, thus some of the countries with largest percentage of the elderly will experience more challenges than others. A common health-related feature of the elderly and the oldest old is they may present more than one aliment associated with their health. The common diseases among the geriatric population are cancer, cardiovascular diseases, stroke, diabetes, infections, chronic pulmonary diseases, and Alzheimer's, while musculoskeletal complaints, is the globally accepted cause of disability among the older populations. Increasing age, diseases, and disabilities will therefore cast an additional pressure on healthcare and long-term care services for the oldest-old and the elderly, and the preparedness to tackle such concerns may still require coherent planning.

Keywords Aging · Age-related diseases · Disabilities · Healthcare · COVID-19

2.1 Introduction

Aging is a global concern that ought to be considered as a trademark feature of societal progression and also as a burden of diseases and disabilities. Aging poses challenges in terms of healthcare and maintaining optimal health for the growing demographic regions globally [1]. According to the World Population Aging 2020 Highlights [2], there were approximately 700 million people in the world who were 65 years or above in 2020, and this number is projected to rise to almost 1.5 billion by 2050. The proportion of people nationwide over the age of 65 years has increased in the previous 10 years and is projected to keep rising for the next 20 years due to better and longer life expectancies. Some countries are aging considerably faster than any other nation [1].

B. Khandelwal (✉) · C. Gupta
Sikkim Manipal Institute of Medical Sciences, Sikkim Manipal University, Gangtok, Sikkim, India
e-mail: bidita.k@smims.smu.edu.in

P. Lama (ed.), *The Ageing Population*,
https://doi.org/10.1007/978-981-99-5772-9_2

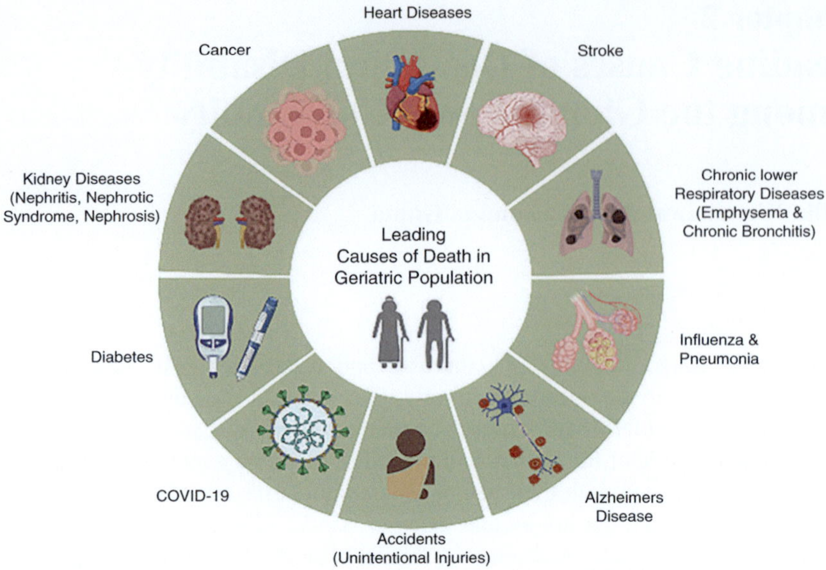

Fig. 2.1 Leading causes of death in the aging population (created with Bio Render)

A large proportion of fatalities among the geriatric population are brought on by long-term chronic diseases (Fig. 2.1) such as cancer, cardiovascular disease (CVD), stroke, diabetes, infections, chronic pulmonary diseases, and Alzheimer's disease [3]. COVID-19 remains to be critical cause of high mortality among older population [4]. Additionally, it has been demonstrated that multimorbidity and geriatric disorders such as frailty augment their risk [5]. In terms of public health, older adults Osteoarthritis account for 23% of the global burden of disease [6]. Multiple medical problems in older adults have been attributed to increased use of drugs, polypharmacy manipulations, adverse health outcomes higher healthcare expenses, disability, a poor quality of life, and even mortality.

The identification of co-occurrence is challenged by the fact that multimorbidity research has traditionally focused on counting diseases. However, new research must underscore the significance of considering multiple chronic diseases into account for the assessment of geriatric health [7, 8]. In accordance with the WHO's 2019 Global Health Estimates, noncommunicable diseases (NCDs) have accounted for seven of the leading ten fatalities worldwide. This represents an upsurge from four of the top ten causes in 2000. The updated data covers the years 2000 through 2019 inclusively. The estimates show changes in disease- and injury-related death and disability rates during the last 20 years. As outlined in the plan for the UN Sustainable Development Goals unequivocally demonstrate the necessity of global attention on combating and managing diseases including chronic respiratory diseases, cancer, diabetes, and cardiovascular diseases as well as injuries, in every sector of the World [9]. The majority of NCD cases affect individuals between the ages of 50 and 69 years. Excepting for HIV/AIDS and tuberculosis, which are major causes of death in some still developing nations, the diversity in causes of death among people aged 50–69 years is far lower globally than it is for younger age groups [10].

NCDs still tend to be the prime factor causing fatality in the oldest demographic, 70 years and above. The other causes of mortality, which include Alzheimer's/dementia and even infections caused by diarrhea, are becoming more common. The majority of low-income countries persist to be plagued by bowel-related illnesses as one of the few leading causes of death for those aged 70 years and older, regardless of the fact that they are less common in nations with better revenues [10]. Considering the global trend of aging, there is still an inadequate amount of information on the health and time indicators of health in the geriatric population. There is certain evidence over the past century that chronic diseases have surpassed infectious diseases as the leading cause of healthcare expenditures and that aging is closely associated with practically all chronic diseases. Population-based research projects on age-related chronic diseases are currently being undertaken with a new focus as many medical professionals and policymakers have grown to comprehend these issues, and realize the cost of care.

2.2 Causes and Risk Factors for Death and Disability among the Aging Community

Understanding the definitions of the term "cause of death" and "risk factors" for an aging population is crucial. In accordance with the "epidemiological framework of the Global Burden of Disease" studies, every mortality has an identifiable cause. This is distinct from deaths and disability that occurred as a result of risk factors. These fatalities represent an estimate of the number of deaths that would be prevented if risk variables to which a given group is exposed were obliterated (for example, tobacco or smoking) or lowered to an ideal threshold. Risks can be categorized into four main groups related to behavior, the environment, occupation, and metabolism [10, 11]. High blood pressure, smoking, obesity, high blood sugar, and cholesterol, use of alcohol and others are the multi-risk factors for death in the population above 65 years of age [11].

There are several diagnostic and management challenges in the elderly due to the following reasons:

- They present atypically and with nonspecific symptoms.
- May have rapid deterioration which may be fatal.
- Health, social, and financial sequelae of delayed diagnosis are enormous.
- Comorbidities make the condition more complex.
- Lack of physiological reserve leads to significant disability and a non-fatal condition to be fatal.

2.3 Neurological Diseases in Elderly

Neurological disorders are common among the geriatric population, involving 5–55% of those aged 55 years and above. They pose a significant risk for unfavorable health outcomes, such as mortality, long-term substantial disability, confinement, and acute hospitalization. Dementia, Alzheimer's disease, Parkinson's disease, and stroke are other leading causes of death and disability in older adults [12].

Dementia and Alzheimer's Disease Dementia, a type of cognitive impairment, is significantly related to old age. Nearly, 10% of people over the age of 65 years' experience some form of memory loss. Alzheimer's disease (AD) and vascular dementia (VaD) are the two most frequent causes of dementia. Nearly 50% of those over 85 years have dementia, which becomes more prevalent as people become older. As a gradual condition, dementia eventually leaves subjects severely disabled and totally dependent on caretakers [13]. Another characteristic of dementia is the impairment and deterioration of cells that express acetylcholine (ACh) receptors, often known as "cholinergic neurons" (AChN), which happens progressively as people age and the disease progresses leading to behavioral and cognitive disorders, particularly learning impairments [14].

The most prevalent type of dementia in elderly individuals is Alzheimer's disease (AD). Aging is one of the main risk factors for Alzheimer's, which is a progressively incurable neurological disease. Two main clinical subgroups of AD can be distinguished: familial AD (FAD), which typically has an early onset (before 65 years of age) and APP, PSEN1, and PSEN2 gene alterations are suggested to be the reasons behind the disease. Almost 50% of the FAD cases observed are caused by this genetic triad [15]. In humans, these alterations to genes do not happen frequently. As a result, the majority (90%) of AD cases are sporadic (SAD) and primarily affect those over 65 years. Finding a SAD causal component is particularly challenging due to the complex nature of the condition [16].

Clinical signs and symptoms are used to make the diagnosis of suspected AD, but a definitive diagnosis cannot be made until after death, and a histopathologic examination of the brain reveals the disease's distinctive lesions, particularly β plaques and neurofibrillary tangles (NFTs) made of the hyperphosphorylated protein. Even though many AD patients can die from other causes, the disease is a life-limiting one. Lack of ability to swallow can be a symptom of Alzheimer's disease. Aspiration (food inhaled into the lungs) may result from this, which increases the risk of chest infections. Patients with this condition frequently develop difficulties in eating and lose their appetite over time. The need for palliative care for those with Alzheimer's disease is becoming more widely recognized. Support by both family and the hospital staff is part of the healthcare [17].

Parkinson's Disease Parkinson's disease (PD) is the second most common age-associated neurological condition after Alzheimer's Disease. PD is characterized by bradykinesia, stiffness, and resting tremor. Due to age-related neurological changes, the proportion of people with PD continues to expand over time. Dopaminergic (DA) neuron loss in the substantia nigra (SN), a defining feature of PD, is caused by the formation of Lewy bodies (LBs), which are brought on by malfunctions in the ubiquitin-proteasome system and abnormalities in the proteins alpha-synuclein (ASN) and Parkin. Increased ROS generation brought on by mitochondrial failure, together with a reduction in mitophagy, neurological inflammation, and depletion of neurotrophic factors, or mutations in PARK genes, are some of the molecular characteristics of Parkinson's disease [18]. The major risk factor for acquiring Parkinson's disease is advanced age is 1% of those over 60 years and 5% of people over 85 years, affected by PD [19].

Stroke WHO defines a stroke as "rapidly developing clinical signs of focal (or global) disturbance of cerebral function, lasting more than 24 h or leading to death, with no apparent cause other than that of vascular origin" [20]. Strokes are classified into two categories: ischemic and hemorrhagic, each with its own pathophysiology. Stroke is the third most common cause of death and the main cause of disability among older adults in Indian rural, semi-urban, and urban settings, Stroke is also a leading cause of disability and premature mortality [21]. Age is the unmodifiable risk factor of stroke, and the rate of stroke is predicted to double every 10 years beyond the age of 55 years and can impact up to 5% of adults over the age of 65 years. It has been established that several modifiable risk factors are more prevalent in the elderly, notably high blood pressure, atrial fibrillation, hyperhomocysteinemia and inflammation. It is evidenced that age-associated white matter (WM) alterations raise the likelihood of stroke, poststroke death, and ischemia injury following stroke. Age-related changes in the brain's mitochondrial and antioxidant pathways lead to higher levels of oxidative stress (OS) and increase the damage to the white matter (WM) caused by ischemia [22].

2.4 Cancer and Aging

There has been an 11-fold increase in malignancies after the age of 65 years, compared to younger adults [23], and 80% of the diagnosis are often made after the age of 55 years [24]. The median age of diagnosis of many cancers is beyond 60 years [25]. As there is a relative decrease in death from other causes, the impact of cancer mortality increases. Treating elderly patients with cancer according to their physical health and their preferences, involves geriatric assessment, geriatric interventions only if necessary, and then individually best-suited therapy is administered, though there is a dearth of evidence.

An alarming increase in the global cancer burden was forecasted in the report released in 2012 by WHO, and International Cancer Research Institute, suggests the aging population is going to bear the brunt of it *(World Cancer Report, 2014).* This increase also reflects as an increased cancer-related mortality as there are several limitations in the screening guidelines, diagnosis, and treatment modalities in the elderly group. Cancer biology in the aging population is different. Delayed diagnosis leads to a worse prognosis despite slow progression of the disease. Fear of the disease and its outcome often leads to concealment of the symptoms. Geriatric oncology is thus, the need of the hour so that age-specific challenges can be addressed. Differences in the tumor extrinsic factors, intrinsic factors, and host factors, between various age groups, not only influence the outcome, but also the effectiveness of treatment.

Breast Cancer Commonest cancer in women worldwide, with a mean age at diagnosis of 62 years, is the incidence of breast cancer [26]. Elderly women have intermediate to low-grade luminal tumors and less incidence of triple-negative sub-type. Advancement in the diagnosis and management have led to decrease in the mortality due to breast cancers, however, country-specific spending on health systems may differ, and some may still die due to lack of adequate treatment toward breast cancer.

Lung Cancer Lung cancer is a common malignancy, with a peak incidence between the ages of 59 and 75 years and one of the leading causes of cancer-related deaths [27]. Strongest association has been observed with cigarette smoking [28]. Advanced stage of the disease with metastasis is often seen on diagnosis. Cough, weight loss, and dyspnea are the common presenting symptoms. In addition to the histopathological understandings, the age-related functional status needs to be considered in deciding the modality of treatment. Aged patients are less likely to receive appropriate therapy compared with young patients due to lack of evidence unfortunately.

Prostate Cancer Prostate cancer is the second most commonly diagnosed cancer and the fifth leading cause of cancer death among men worldwide [29]. At diagnosis, 60% of patients are aged 65 years and older. Prostate cancer mortality in men aged 70 years and older is expected to almost double between 2018 and 2030 [30], especially in the elderly population age range aggressive form of cancer is encountered, as interpretation of prostate-specific antigen is hampered in this age group and thus, the need for age-based reference values and other markers is advisable.

Melanoma The median age of diagnosis of melanoma is 65 years, and it is the fifth most common malignancy with age-adjusted incidence. With an average annual increase of 1.5% in incidence rates between 2008 and 2017, it is predicted that 2.2% of men and women will be diagnosed with melanoma of the skin at some point during their lifetime [31]. Adverse prognostic histological characteristics lead to chronic disability and tumor burden in the aging population.

Colorectal Cancer Colorectal cancer is diagnosed at a median age of around 70 years, with marked biological differences in comparison with younger patients. Higher mortality and morbidity with advancing age is observed with relevance to this cancer sub-type and is especially common in the North American, European, and Oceanic regions.

2.5 Cardiovascular Disease in the Elderly

Structural and functional changes occur in the heart as physiological aging process which leads to an increased risk of Cardiovascular Diseases (CVD) and high burden in the elderly group. Diastolic and systolic dysfunction leads to contractility and relaxation impairments and electrical dysfunction leads to arrhythmias, i.e., rhythmic disturbances. Structural changes lead to vascular remodeling and vascular stiffness which in addition to increasing the risk for cardiac events, also results in cerebrovascular events, peripheral vascular disease, and cognitive dysfunction. Several factors, such as increased oxidative stress, inflammation, apoptosis, and myocardial degeneration further contribute to the deteriorating heart [32]. CVD is the commonest cause of death in the elderly population [33]. Despite the decline in mortality due to CVD in the aging population, it still tops the list and with the associated morbidity burdening the healthcare costs. Above 80 years, females have a greater incidence of CVD. Decline in hormones is thought to be responsible for this gender difference in old age, though there has been no established benefit of hormone replacement [34].

Challenges in Diagnosing and Managing CVD in Elderly Overdiagnosis as well as underdiagnosis of CVD is observed in this special group of population due to several reasons.

- Atypical presentation with vague and nonspecific symptoms like breathlessness, loss of consciousness, upper abdominal pain, or fatigue.
- Symptoms masked by other co-morbidities.
- Silent Myocardial infarction, i.e., Painless heart attack is common in those with Diabetes Mellitus.
- Absent ECG changes or difficult to interpret ECG due to co-existing LBBB.
- Increased Troponin levels due to other causes being misinterpreted as acute coronary events, complaints are ignored as part of normal aging.
- Aggressive treatment and secondary prevention are not offered due to this population being not included in most of the clinical trials and lack of effectiveness of treatment in this group.

Heart Failure Is the commonest CVD in the elderly with 1 in 10 suffering from it above 65 years of age. Overall prevalence is 3–20 cases per 1000 population, which after 45 years doubles with each decade [35]. The mortality and morbidity are higher in those with systolic heart failure, in male Caucasian [36]. Medical management is the same as that in the younger group but support clinics led by trained nurses have an enormous impact in reducing repeated hospitalizations, and improving the adherence to treatment as well as in educating the elderly population, which is an expensive process.

Coronary Artery Disease Age is an independent risk factor for coronary artery disease and the incidence increases with advancing age. Modifiable risk factors are similar to those in the young but often not targeted in the elderly population.

Atypical presentation leading to delayed diagnosis as well as misdiagnosis is often responsible for mortality. Forty-two percent of myocardial infarctions (MI) were asymptomatic in men aged 75–84 years compared with only 18% in men between 45 and 54 years, as per the Framingham study [35].

Arrhythmias Irregularity in the heart rhythm often presents as presyncope, giddiness, recurrent falls, fatigue, worsening of breathlessness, or palpitations. Lone atrial fibrillation and ventricular arrhythmias in the aged can cause sudden cardiac death. Bradyarrhythmia may be a precursor or a presentation of various heart blocks [36].

Hypertension Is one of the most important risk factors for CVD. The prevalence in those above 65 years is reported to be as high as 60–80%. With advancing age, the systolic pressure increases gradually while the diastolic pressure falls slightly, thus leading to a widened pulse pressure. Isolated systolic hypertension is also a risk factor for stroke, Ischemic heart disease, peripheral vascular disease, and congestive heart failure, as well as a predictor of adverse cardiovascular outcomes that need to be addressed and not be neglected as part of normal aging. Treatment guidelines and targets to be achieved are well defined. Postural hypotension needs to be avoided and so the use of drugs should be judicious [37, 38].

2.6 Respiratory Diseases in the Elderly

The complex biologic process of aging causes irreversible structural and functional changes in the respiratory system resulting in respiratory diseases of varying severity. At one end of the severity spectrum lies mild cough and rhinitis which are commonly occuring respiratory illnessess, and at the other end are life threatening respiratory illnessess like influenza and lung cancer. In between, bronchial asthma and chronic obstructive airway diseases having a chronic course with exacerbations, leading to high morbidity and even mortality [39].

With advancing age, there is a reduction in the cartilaginous portion of the nasal passage resulting in reduced patency. Air trapping in the alveoli and mismatched ventilation-perfusion due to the loss of collagen network, leads to ineffective gas exchange [39, 40]. The contractility of the respiratory muscles including intercostal, diaphragmatic, and chest wall, decreases as part of aging, leading to ineffective cough, decreased airway expansion, and inadequate mucociliary clearance [40]. Environmental factors like smoking and pollution along with the genetic component decide the rate at which this deterioration takes place. Atypical manifestations often complicated by co-morbidities and false-positive tests lead to a diagnostic challenge in elderly group of patients with respiratory diseases.

Bronchial Asthma Asthma is a chronic inflammatory condition characterized by reversible airway obstruction resulting in recurring and variable symptoms of cough, wheeze, shortness of breath, and chest tightness. At 65 years and above, Asthma has a lifetime prevalence of 13% while self-reported asthma in this age group is 6.8%

[41, 42]. Prolonged hospital stays, repeated severe exacerbations and increased mortality in the elderly population makes it an important healthcare issue [43]. Often the symptoms are subtle and easily missed or neglected. Any elderly person presenting with features of dyspnea, cough, chest tightness, and wheeze, should always be investigated for asthma.

Interpretation of the spirometry results in the elderly must consider the age-adjusted values. Frailty and cognitive decline often lead to difficulty in performing accurate spirometry. Inconsistent results on provocation tests and contradictory results in fractional exhaled nitrous oxide make the diagnosis even more challenging. Age-customized management as per the existing guidelines by Global Initiative for Asthma, should be instituted with the primary aim to reduce mortality and morbidity [42–44].

Chronic Obstructive Pulmonary Disease COPD is defined by the Global Initiative for Chronic Obstructive Lung Disease (GOLD) as "A common, preventable, and treatable disease that is characterized by persistent respiratory symptoms and airflow limitation that is due to airway and/or alveolar abnormalities usually caused by significant exposure to noxious particles or gases" [44]. It ranks among the top five causes of death worldwide along with high burden of disability.

Respiratory Infections Features of respiratory infections like dry cough or cough with expectoration, difficulty in breathing, fever, or chest pain are very common presentations in the elderly. The part of airway involved determines the severity and management approach. **Influenza** is the severest form of viral respiratory tract infection often leading to systemic illness and sepsis with bacterial superinfection. Due to the other co-existing illness, the mortality due to this is high. **Pneumonia** is an infection of one or both of the lungs caused by bacteria, viruses, or fungi and a leading cause of mortality in the elderly. Prompt treatment with antimicrobials, supplemental oxygen, optimal fluid and chest physiotherapy reduces mortality. In all above 65 years of age, vaccination against pneumococcal given once and against influenza given annually, reduces the severity to a great extent [43, 44].

2.7 Musculoskeletal Impairments and Incidence of Falls with Age

Wrinkles and gray hair are inevitable age-related changes but changes in muscles, bones and joints associated with aging can be reduced or even reversed by physical activity. Approximately 1.71 billion people have musculoskeletal (MSK) impairments which are the leading contributor to disability worldwide. MSK diseases significantly limit mobility and dexterity, leading to reduced quality of life and decreased ability to participate in society [32]. According to the World Health Organization, one of the major disabling conditions among the elderly population is musculoskeletal (MSK) disorders [33].

Osteoarthritis Osteoarthritis (OA) is the most common joint disorder in older patients, causing a massive burden of morbidity, dependency, and pain. It is a disorder of the dynamic repair process of synovial joints, leading to loss of articular cartilage (joint space narrowing), vascular congestion, new growth of bone and cartilage (osteophytes), and capsular fibrosis. Increasing age and females are at higher risk of developing OA. Obesity and trauma are potentially avoidable risk factors that can minimize early incidence of OA in younger age groups. Severe OA can contribute to postural instability, falls, and thus deaths occasionally. Nonpharmacological treatment including exercise, hot/cold packs, weight reduction, and comfortable footwear; pharmacological treatment with local and oral analgesics and intra-articular steroids and surgical treatment including joint replacement are beneficial [45–47].

Osteoporosis and Osteopenia It is normal loss of bone density with aging. Severe weakening of bone density is classified as Osteoporosis particularly common in the elderly and is associated with an increased rate of bone fractures, while osteopenia is not. Bone density screening is recommended for women aged above 65 years [47]. Although the prevalence of fractures in men increases by age 85 years, the value of osteoporosis screening for men has not been clearly demonstrated. The effectiveness and safety of calcium and vitamin D supplementation to prevent fractures remains controversial.

Falls Falls are the second leading cause of unintentional injury and death worldwide. Each year an estimated 684,000 individuals die from falls and adults older than 60 years of age and females are at a greater risk. This risk level may be in part due to physical, sensory, and cognitive decline associated with aging, in combination with environment, for example, snowfall and rain, and the unavoidable changes in the surrounding surfaces which may not be well suited for an aging population [48]. Non-fatal falls or those which do not result in significant physical injury, can still have a psychological impact increasing self-restriction of activities. Elderly persons who survive a fall experience may undergo significant morbidity in terms of prolonged hospitalization and greater functional decline in activities of daily living [45–48]. Falls are typically multifactorial in nature, the result of interactions among various impairments, especially those affecting gait, balance, vision, accommodation, hearing, behavior, and the surrounding environment. Vestibular disorder, orthostatic hypotension, cardiac arrhythmias, and unstable posture may all be contributing causes toward unexpected falls. Evidence suggest that physical activity makes bones stronger and help slow the rate of bone loss, increases muscle mass and strength, improves balance and coordination, and thus reduce the risk of fall and helps in maintaining joint flexibility and functions [49].

2.8 Diabetes Mellitus and Aging

Adults aged 65 years and above have a very high prevalence of diabetes [49]. Type 2 diabetes is more common than Type 1, both in the adult population as well as in the geriatric age group [45]. In adults 65 years and above, the combined prevalence of diagnosed and undiagnosed prediabetes and diabetes ranges from 50% to 80% [50]. The projected increase in diabetic patients aged 65 years and above is set to increase fourfold by 2050 [51]. One of the strongest risk factors for cardiovascular disease at the age of 85 years is diabetes. Peripheral neuropathy and peripheral arterial disease results in diabetic foot which leads to disabilities and is the major cause for amputations due to uncontrolled diabetes. Longer duration of the disease in older diabetics may have more complications like cardiomyopathy, chronic kidney disease (CKD), and retinopathy [45], and diabetic cardiomyopathy and CKD are the major cause of mortality in diabetic patients. Management of diabetes needs to be tailored based on the comorbidities and the associated diabetic complications. Proper assessment of a patient followed by implementation of treatment by judicious selection of medicines from the wide range of antidiabetic medications available is essential. This will ensure cost effective management with optimal glycemic control and least risk of hypoglycemia [51, 52].

2.9 COVID-19 and the Impact on Elderly

Since the novel Coronavirus-19 (COVID-19) pandemic first appeared in early 2020 elderly population are more susceptible to it and has serious complications due to social, as well as, medical reasons. In comparison to young adults, unvaccinated elderly people with disabilities, and often with a low threshold for immunological diseases, and/or with certain underlying medical disorders, were more likely to require hospitalization, critical care, ventilation and many eventually died from COVID-19. The majority of COVID-19 deaths involve adults over 65 years of age [53]. Initial studies from China indicated a rise in disease severity and death in individuals aged 60 years and older, and records from European studies also showed a comparable trend, with mortality predicted to be as high as 10% in those 70 years and older, relative to 1% in young adults [47]. The respiratory disease COVID-19 is brought on by the infection caused by Severe Acute Respiratory Syndrome coronavirus (SARS-CoV-2). COVID-19 is primarily spread by respiratory droplets and through direct contact [54].

Elderly patients with respiratory infections may present a typical characteristic feature, such as exhaustion, anorexia, and dementia, contrary to fever and a productive cough, which leads to delayed diagnosis. Age is thus a significant risk factor for mortality among other various age groups [55]. Older adults with reduced levels of antiviral antibodies and C-reactive protein (CRP), a powerful marker of inflammation, had more lung lobes affected, required ventilatory assistance, and were more likely to experience blood-gas irregularities [53]. One of the difficulties with the COVID-19 pandemic was the presence of indeterminate organ malfunction in the elderly, as most of the death reported were from congestive heart failure and sepsis yet the infected showed no symptoms of pneumonia. Additionally, elderly may even have neurological and sensory impairments that made it challenging for them to understand and adhere to safety procedures [52, 53].

2.10 Hearing Impairment with Aging

Hearing impairment is seen in 50% of adults over 85 years of age [56]. Decline in hearing that commonly occurs with age is known as presbyacusis, which is due to degeneration of both the sensory peripheral (cochlea), accountable for 75% of the clinical features, and central (neural) components of the auditory system. Greater hearing loss and earlier onset are experienced in men compared to women [57]. Despite its prevalence and morbidity, hearing loss is often underrecognized due to its slow onset and progression and thus undertreated. Cost and social stigma are major factors in the diagnosis and management of hearing loss [58].

Mild hearing loss can impair speech processing if the speech is fast, or when many persons talk together. Increased social isolation explains the associations between hearing loss and depression, cognitive decline, and reduced quality of life in the elderlies [56]. The use of hearing aids can bring functional improvement in the elderly but the use of it is limited [58] as many may report unsatisfactory results and non-coverage by insurance agencies are other factors responsible for low usage. The American Academy of Family Physicians recommends screening older adults above 60 years for hearing loss during their periodic health examinations by readily available screening tests like whispered voice test, audioscopy, or audiometry [57]. Patient-centered treatment with hearing aids, assistive listening devices, and surgical implants are some of the options available for patients with progressive hearing loss. The onset of age-related hearing loss can be delayed by avoiding risk factors and noise. Folic acid (800 mcg daily) and a high intake of omega-3 fatty acids may additionally slow hearing decline [58, 59].

2.11 Visual Impairment with Aging

Healthy aging includes healthy eyes, as impairment of visual function leads to disability compromising functional tasks. Aging leads to reduction in visual acuity, visual field, color vision, dark adaptation, and power of accommodation often leading to presbyopia. Due to the decreased functioning of other sensory systems, related to age or to medical diseases like deafness, diabetes mellitus, and neuropathy, vision plays an enhanced role in gaining critical information from the surroundings to remain steady. Loss of vision among the elderly is thus a major healthcare problem. Approximately one in three elderly persons have some form of visual impairment by the age of 65 years [60], and it may be associated with functional disability and an increased risk for depression [61]. Some of the Common causes of visual impairments are as follows.

Age-Related Macular Degeneration Age-related macular degeneration (AMD) is the leading cause of visual loss in people over 65 years of age and is characterized by degeneration of the macula, the area of the retina responsible for central vision. Risk factors for AMD include advancing age, family history of AMD, cardiovascular risk factors such as hypertension and cigarette smoking. Diet rich in carotenoids may lower the risk of developing advanced or exudative AMD [62].

Glaucoma Glaucoma comprises a group of disorders characterized by glaucomatous optic nerve damage and visual field loss. Approximately one million individuals over 65 years of age have glaucoma-associated visual impairment, and approximately 75% of persons who are legally blind because of glaucoma are over the age of 65 years [62]. The most prevalent form of glaucoma is the open-angle sub-type which usually responds to medications. Surgical interventions like laser or incisional techniques are reserved for those who do not respond satisfactorily.

Cataract It is the commonest cause of visual impairment in thc elderly and the most common cause of blindness worldwide [63, 64]. The prevalence of cataract increases with age from less than 5% in persons under 65 years of age to approximately 50% in those above 75 years of age and older [65]. Exposure to ultraviolet light may contribute to the progression of cataract formation.

Diabetic Retinopathy A significant cause of vision morbidity in the elderly population and the leading cause of new blindness among middle aged is due to diabetic retinopathy. The prevalence of diabetic retinopathy rises with increasing duration of diabetes. But, significant diabetic retinopathy may be observed in the elderly at diagnosis or during the first few years of diabetes [66]. Optimal glycemic control of DM and management of co-morbidities like hypertension, hyperlipidemia, fluid overload states, and renal failure reduces the development and retards the progression of diabetic retinopathy [67].

Simple tips that help to maintain healthy eyes are protection from UV radiation of the Sun, avoidance of smoking, physically active life, healthy diet, good control of co-existing conditions like DM or HTN, and avoidance of eye strain.

2.12 Miscellaneous Causes of Disability in the Elderly

Geriatric Syndromes These syndromes result from the cumulative, synergistic effects of discrete diseases on multiple organ systems when the body fails to compensate, and the effects are finally expressed as a single symptom [68]. Geriatric Syndromes (GSs) commonly include cognitive decline, depression, delirium, gait disorders, falls, sarcopenia, sleep disorders, urinary incontinence, pressure ulcers, sensory impairments, chronic pain, and frailty.

Frailty It can be defined as a transitional phase between successful aging and disability. Frailty is a vulnerable state characterized by the accumulation of impairments in several interdependent systems, a decrease in physiological reserves, and the failure of homeostatic mechanisms in response to low-level stress [69]. Women have a higher prevalence and risk of frailty but have a lower risk of mortality, the so-called male–female healthy survival paradox [69], which is attributed to a combination of behavioral, social, and biological factors. Protective effect of estrogen against female mortality, particularly in terms of CVD is the most evidenced. Frailty is special vulnerability to stressors and is presented as weakness, slowness, exhaustion, and weight loss and may be as high as 38% in people aged 85–89 years [70]. Frailty predicts future disability, risk of fall, hospitalization, and poor surgical outcomes and so needs to be assessed and addressed, especially amongst the oldest old.

Continence Urinary incontinence has been reported in 30% of women over the age of 65 years and half of older adults in nursing care facilities [71]. Stress and functional incontinence, as well as overactive bladder, are commonly responsible for it in women. Urinary incontinence negatively impacts the quality of life, and medical treatment is often not well tolerated [72].

Immuno-Senescence Age-related changes in the immune system lead to diminished cellular and humoral immunity and thus, increased susceptibility to infection and delayed wound healing [73]. Chronic inflammation mediates depression of B-cell function in addition to reduction in T-cell generation and activation. Dysfunctional innate immunity contributes to increased severity of infections in the elderly population [74]. Reduced effectiveness of vaccines in this group is a matter of concern, but effectiveness of a higher dose has not shown enhanced protection [75].

Conclusion Population of the world is progressively aging. This transition renders positive outcomes and concerns. Most of us would want to live long and healthy lives, and this would increase the proportion of elderly people who can be an important resource in terms of economic manpower, society, culture, and family support. However, elderly communities are also linked to an upsurge in the need for medical care and social assistance, and would increase the expenditure for healthcare.

Competing Interest The authors have no conflicts of interest to declare that are relevant to the content of this chapter.

References

1. United Nations, Department of Economic and Social Affairs, Population Division. World Population Ageing 2019 (ST/ESA/SER.A/444) 2020.
2. Population Division | [Internet]. [cited 2023 Jun 10]. Available from: https://www.un.org/development/desa/pd/
3. Gorina Y, Hoyert D, Lentzner H, Goulding M. Trends in causes of death among older persons in the United States: (620642007-001) [Internet]. 2005 [cited 2023 Jun 8]. Available from: http://doi.apa.org/get-pe-doi.cfm?doi=10.1037/e620642007-001
4. COVID-19 Risks and Information for Older Adults [Internet]. Centers for Disease Control and Prevention. 2023 [cited 2023 Jun 10]. Available from: https://www.cdc.gov/aging/covid19/index.html
5. Beard JR, Bloom DE. Towards a comprehensive public health response to population ageing. Lancet. 2015;385(9968):658–61.
6. Prince MJ, Wu F, Guo Y, Gutierrez Robledo LM, O'Donnell M, Sullivan R, et al. The burden of disease in older people and implications for health policy and practice. Lancet. 2015;385(9967):549–62.
7. Arokiasamy P, Uttamacharya U, Jain K, Biritwum RB, Yawson AE, Wu F, et al. The impact of multimorbidity on adult physical and mental health in low- and middle-income countries: what does the study on global ageing and adult health (SAGE) reveal? BMC Med. 2015;13(1):178.
8. Marengoni A, Angleman S, Melis R, Mangialasche F, Karp A, Garmen A, et al. Aging with multimorbidity: a systematic review of the literature. Ageing Res Rev. 2011;10(4):430–9.
9. WHO reveals leading causes of death and disability worldwide: 2000–2019 [Internet]. [cited 2023 Jun 7]. Available from: https://www.who.int/news/item/09-12-2020-who-reveals-leading-causes-of-death-and-disability-worldwide-2000-2019
10. Ritchie H, Spooner F, Roser M. Causes of death. Our World in Data [Internet]. 2018 Feb 14 [cited 2023 Jun 10]; Available from: https://ourworldindata.org/causes-of-death
11. Gakidou E, Afshin A, Abajobir AA, Abate KH, Abbafati C, Abbas KM, et al. Global, regional, and national comparative risk assessment of 84 behavioural, environmental and occupational, and metabolic risks or clusters of risks, 1990–2016: a systematic analysis for the Global Burden of Disease Study 2016. Lancet. 2017;390(10100):1345–422.
12. Callixte KT, Clet TB, Jacques D, Faustin Y, François DJ, Maturin TT. The pattern of neurological diseases in elderly people in outpatient consultations in Sub-Saharan Africa. BMC Res Notes. 2015;8:159.
13. Dorszewska J, Prendecki M, Oczkowska A, Dezor M, Kozubski W. Molecular basis of familial and sporadic Alzheimer's disease. Curr Alzheimer Res. 2016;13(9):952–63.
14. Tata AM, Velluto L, D'Angelo C, Reale M. Cholinergic system dysfunction and neurodegenerative diseases: cause or effect? CNS Neurol Disord Drug Targets. 2014;13(7):1294–303.
15. Selkoe DJ. Alzheimer's disease: genes, proteins, and therapy. Physiol Rev. 2001;81(2):741–66.
16. Alonso Vilatela ME, López-López M, Yescas-Gómez P. Genetics of Alzheimer's disease. Arch Med Res. 2012;43(8):622–31.
17. Alzheimer's disease [Internet]. nhs.uk. 2018 [cited 2023 Jun 11]. Available from: https://www.nhs.uk/conditions/alzheimers-disease/
18. Reeve A, Simcox E, Turnbull D. Ageing and Parkinson's disease: why is advancing age the biggest risk factor? Ageing Res Rev. 2014;14:19–30.
19. Raver SM, Lin SC. Basal forebrain motivational salience signal enhances cortical processing and decision speed. Front Behav Neurosci [Internet]. 2015 [cited 2023 Jun 10]; 9. Available from: https://www.frontiersin.org/articles/10.3389/fnbeh.2015.00277
20. Aho K, Harmsen P, Hatano S, Marquardsen J, Smirnov VE, Strasser T. Cerebrovascular disease in the community: results of a WHO collaborative study. Bull World Health Organ. 1980;58(1):113–30.
21. Why are strokes more common in the elderly? [Internet]. 2015 [cited 2023 Jun 11]. Available from: https://kauveryhospital.com/blog/neurology/stroke-in-the-elderly/

22. Oksala NKJ, Oksala A, Pohjasvaara T, Vataja R, Kaste M, Karhunen PJ, et al. Age related white matter changes predict stroke death in long term follow-up. J Neurol Neurosurg Psychiatry. 2009;80(7):762–6.
23. Yancik R. Cancer burden in the aged: an epidemiologic and demographic overview. Cancer. 1997;80:1273–83.
24. Pallis AG, Fortpied C, Wedding U, et al. EORTC elderly task force position paper: approach to the older cancer patient. Eur J Cancer. 2010;46:1502–13.
25. Arnold M, Karim-Kos HE, Coebergh JW, et al. Recent trends in incidence of five common cancers in 26 European countries since 1988: analysis of the European Cancer Observatory. Eur J Cancer. 2015;51:1164–87.
26. Bidoli E, Virdone S, Hamdi-Cherif M, et al. Worldwide age at onset of female breast cancer: a 25-year population-based cancer registry study. Sci Rep. 2019;9:14.
27. O'Rourke MA, Feussner JR, Feigl P, Laszlo J. Age trends of lung cancer stage at diagnosis. JAMA. 1987;258:921–6.
28. Alberg A, Samet J. Epidemiology of lung cancer. Chest. 2003;123:21–49S.
29. Bray F, Ferlay J, Soerjomataram I, Siegel RL, Torre LA, Jemal A. Global cancer statistics 2018: GLOBOCAN estimates of incidence and mortality worldwide for 36 cancers in 185 countries. CA Cancer J Clin. 2018;68:394–424.
30. Boyle HJ, Alibhai S, Decoster L, et al. Updated recommendations of the International Society of Geriatric Oncology on prostate cancer management in older patients. Eur J Cancer. 2019;116: 116–36.
31. Cancer stat facts: melanoma of the skin. 2020. https://seer.cancer.gov/statfacts/html/melan.html (accessed Jun 10, 2023).
32. Cieza A, Causey K, Kamenov K, Hanson SW, Chatterji S, Vos T. Global estimates of the need for rehabilitation based on the Global Burden of Disease study 2019: a systematic analysis for the Global Burden of Disease Study 2019. Lancet. 2021;396(10267):2006–17.
33. Curtis AB, Karki R, Hattoum A, Sharma UC. Arrhythmias in patients $\geq$ 80 years of age: pathophysiology, management, and outcomes. J Am Coll Cardiol. 2018;71:2041–57.
34. Yazdanyar A, Newman AB. The burden of cardiovascular disease in the elderly: morbidity, mortality, and costs. Clin Geriatr Med. 2009;25:563–77.
35. National Institute for Health and Care Excellence. Chronic heart failure in adults: management. Clinical guideline CG108. M. 2010. http://www.nice.org.uk/cg108.
36. Vasan RS, Benjamin EJ, Levy D. Prevalence, clinical features and prognosis of diastolic heart failure—an epidemiologic perspective. J Am Coll Cardiol. 1995;26:1565–74.
37. Beckett NS, Peters R, Fletcher AE, et al. HYVET Study Group. Treatment of hypertension in patients 80 years of age or older. N Engl J Med. 2008;358:1887–98.
38. National Institute for Health and Care Excellence (2011). Hypertension in adults: diagnosis and management. Clinical guideline.
39. Gillooly M, Lamb D. Airspace size in lungs of lifelong non-smokers: effect of age and sex. Thorax. 1993;48:39–43.
40. Estenne M, Yernault J, De Troyer A. Rib cage and diaphragm-abdomen compliance in humans: effects of age and posture. J Appl Physiol. 1985;1985(59):1842–8.
41. Moorman JE, Rudd RA, Johnson CA, et al. National surveillance for asthma—United States, 1980–2004. MMWR Surveill Summ. 2007;56:1–54.
42. Parameswaran K, Hildreth AJ, Chadha D, et al. Asthma in the elderly: underperceived, underdiagnosed and undertreated; a community survey. Respir Med. 1998;92:573–7.
43. Tsai C, Lee W, Hanania N, Camargo C. Age-related differences in clinical outcomes for acute asthma in the United States, 2006–2008. J Allergy Clin Immunol. 2012;129:1252–8.e1.
44. Global Initiative for Chronic Obstructive Lung Disease (GOLD). Global strategy for the diagnosis, management and prevention of chronic obstructive pulmonary disease at: https://goldcopd.org/.
45. Woolf AD, Pfleger B. Burden of major musculoskeletal conditions. Bull World Health Organ. 2003;81:646–56.

46. https://www.who.int/news-room/fact-sheets/detail/falls (Accessed 9 Jun 2023).
47. Dunn JE, Rudberg MA, Furner SE, Cassel CK. Mortality, disability, and falls in older persons: the role of underlying disease and disability. Am J Public Health. 1992;82:395–400.
48. Tinetti ME, Doucette J, Claus E, Marottoli R. Risk factors for serious injury during falls by older persons in the community. J Am Geriatr Soc. 1995;43:1214–21.
49. Lee PG, Halter JB. The pathophysiology of hyperglycemia in older adults: clinical considerations. Diabetes Care. 2017;40:444–52.
50. Narayan KM, Boyle JP, Geiss LS, Saaddine JB, Thompson TJ. Impact of recent increase in incidence on future diabetes burden: U.S., 2005-2050. Diabetes Care. 2006;29:2114–6.
51. Odden MC, Shlipak MG, Whitson HE, Katz R, Kearney PM, defilippi C, et al. Risk factors for cardiovascular disease across the spectrum of older age: the cardiovascular health study. Atherosclerosis. 2014;237:336–42.
52. Corriere M, Rooparinesingh N, Kalyani RR. Epidemiology of diabetes and diabetes complications in the elderly: an emerging public health burden. Curr Diab Rep. 2013;13:805–13.
53. Singhal S, Kumar P, Singh S, Saha S, Dey AB. Clinical features and outcomes of COVID-19 in older adults: a systematic review and meta-analysis. BMC Geriatr. 2021;21(1):321.
54. Leung C. Risk factors for predicting mortality in elderly patients with COVID-19: a review of clinical data in China. Mech Ageing Dev. 2020;188:1112.
55. Banerjee D. The impact of Covid-19 pandemic on elderly mental health. Int J Geriatr Psychiatry. 2020;35(12):1466–7.
56. Gates GA, Cobb JL, Linn RT, Rees T, Wolf PA, D'Agostino RB. Central auditory dysfunction, cognitive dysfunction, and dementia in older people. Arch Otolaryngol Head Neck Surg. 1996;122(2):161–7.
57. Popelka MM, Cruickshanks KJ, Wiley TL, Tweed TS, Klein BE, Klein R. Low prevalence of hearing aid use among older adults with hearing loss: the epidemiology of hearing loss study. J Am Geriatr Soc. 1998;46:1075–8.
58. Bagai A, Thavendiranathan P, Detsky AS. Does this patient have hearing impairment? JAMA. 2006;295(4):416–28.
59. Gopinath B, Flood VM, Rochtchina E, McMahon CM, Mitchell P. Consumption of omega-3 fatty acids and fish and risk of age-related hearing loss. Am J Clin Nutr. 2010;92(2):416–4.
60. Ganley JP, Roberts J, eds. Eye conditions and related need for medical care among persons 1–74 years of age, United States, 1971–72. Hyattsville, Md.: U.S. Dept. of Health and Human Services, Public Health Service, National Center for Health Statistics, 1983; DHHS publication no. 83–1678.
61. Shmuely-Dulitzki Y, Rovner BW. Screening for depression in older persons with low vision. Somatic eye symptoms and the geriatric depression scale. Am J Geriatr Psychiatry. 1997;5:216–20.
62. Seddon JM, Ajani UA, Sperduto RD, Hiller R, Blair N, Burton TC, et al. Dietary carotenoids, vitamins A, C, and E, and advanced age-related macular degeneration. Eye Disease Case-Control Study Group. JAMA. 1994;272:1413–20.
63. Pizzarello LD. The dimensions of the problem of eye disease among the elderly. Ophthalmology. 1987;94:1191–5.
64. Foster A, Johnson GJ. Magnitude and causes of blindness in the developing world. Int Ophthalmol. 1990;14:135–40.
65. Kahn HA, Leibowitz HM, Ganley JP, Kini MM, Colton T, Nickerson RS, et al. The Framingham eye study. I. Outline and major prevalence findings. Am J Epidemiol. 1977;106:17–32.
66. Klein R. The epidemiology of diabetic retinopathy: findings from the Wisconsin epidemiologic study of diabetic retinopathy. Int Ophthalmol Clin. 1987;27:230–8.
67. The effect of intensive treatment of diabetes on the development and progression of long-term complications in insulin-dependent diabetes mellitus. The Diabetes Control and Complications Trial Research Group. N Engl J Med. 1993;329:977–86.
68. Inouye SK, Studenski S, Tinetti ME, Kuchel GA. Geriatric syndromes: clinical, research and policy implications of a core geriatric concept. J Am Geriatr Soc. 2007;55:780–91.

69. Clegg A, Young J, Iliffe S, Rikkert MO, Rockwood K. Frailty in elderly people. Lancet. 2013;381(9868):752–62. https://doi.org/10.1016/s0140-6736(12)62167-9.
70. Marinus N, Vigorito C, Giallauria F, Haenen L, Jansegers T, Dendale P, Feys P, Meesen R, Timmermans A, Spildooren J, et al. Frailty is highly prevalent in specific cardiovascular diseases and females, but significantly worsens prognosis in all affected patients: a systematic review. Ageing Res Rev. 2021;66:101233.
71. Bandeen-Roche K, Seplaki CL, Huang J, Buta B, Kalyani RR, Varadhan R, et al. Frailty in older adults: a nationally representative profile in the United States. J Gerontol A Biol Sci Med Sci. 2015;70:1427–34.
72. Shah D, Badlani G. Treatment of overactive bladder and incontinence in the elderly. Rev Urol. 2002;4(Suppl 4):S38–43.
73. Sims J, Browning C, Lundgren-Lindquist B, Kendig H. Urinary incontinence in a community sample of older adults: prevalence and impact on quality of life. Disabil Rehabil. 2011;33: 1389–98.
74. Gould L, Abadir P, Brem H, Carter M, Conner-Kerr T, Davidson J, et al. Chronic wound repair and healing in older adults: current status and future research. J Am Geriatr Soc. 2015;63:427–38.
75. Bandaranayake T, Shaw AC. Host resistance and immune aging. Clin Geriatr Med. 2016;32: 415–32.

Chapter 3
The Societal Impact and Feasibility of Psychosocial Interventions on the Quality of Life of the Elderly

Harshvardhan Sampath and Geeta Soohinda

> No man is an island, Entire of itself...
> *...Therefore, send not to know for whom the bell tolls,*
> *It tolls for thee.*
> *—Devotions Upon Emergent Occasions, "Meditation XVII"*
> *by John Donne.*

Abstract Successful aging should transcend gender, culture, geography, and political discourse. At the same time, it is to be viewed as a personal, subjective process where the elderly dynamically adapt and optimize their reserves and expectations to compensate for the losses of aging. Mental health issues are common among the aged population due to loss of physical functioning, degenerative neurological conditions, and psychosocial factors which negatively impact their quality of life. A holistic, client-driven, bottom up approach addressing the biopsychosocial challenges faced by the elderly is needed. Evidence-based pychosocial interventions forms the backbone of mental care delivery. This chapter explores the plethora of such interventions, especially suited for low-and-middle income countries.

Keywords Aging · Mental health · Psychosocial interventions · Healthcare

3.1 Introduction

John Donne's immortal lines foretell humanity's shared fate and how the intricate web of hope, joy, and suffering unites us all. It serves as a stark reminder to care for the elderly as surely one day the bell will toll for us. Despite the technological advancement of the human race in the twenty-first century, our care for the elderly population has taken a backseat in political and biomedical discourse focusing narrowly on disease and disability [1]. This has led to adding years to life rather

H. Sampath (✉) · G. Soohinda
Central Referral Hospital, Sikkim Manipal Institute of Medical Sciences, Sikkim Manipal University, Gangtok, Sikkim, India
e-mail: harshavardhan.s@smims.smu.edu.in

P. Lama (ed.), *The Ageing Population*,
https://doi.org/10.1007/978-981-99-5772-9_3

than life to those years [2]. Moving beyond senescence to aging needs a paradigm shift in our worldview as global and population aging has a long-lasting impact on this planet [3].

Aging is a continuous biological process, that ticks perpetually in silence, to reveal itself in ominous ways. It is a multidimensional construct that transects core pathways of humanity, physical health, psychological well-being, political ideology, community participation, and environmental stability. These pathways operate on both the micro and macro levels to contribute to the quality of life of the elderly [4]. Unfortunately, narrow and skewed conceptions of aging as a biological process that inevitability leads to infirmity has been an Achilles heel in achieving optimal care for the elderly [5]. From being a participant-driven model, policymakers have myopically focused on a top-down model of care in research and policy-making with watertight compartmentalized geriatric care that is economically unsustainable, especially in low and middle-income countries [6]. It has also reduced the elderly being perceived as a drain on resources rather than as active contributors to society.

This chapter discusses the evidence from diverse fields of research on care planning, implementation, and outcome of holistic care for the elderly guided by various stakeholders to improve the quality of life of the elderly. More importance is given to community-based and public care models rather than institutional and specialized services as they can be implemented in underdeveloped and developing nations.

3.2 Defining Successful Aging

Successful aging is an ideal aspiration that transcends gender, culture, geography, and political discourse. Thus, our understanding of successful aging should incorporate these multiple factors. Objective and elitist models of successful aging have dominated psychosocial discourse for decades. The multi-criteria model of successful aging proposed in the late 1980s was a milestone in aging research [7]. This model was revised later to define "successful agers" as those elderly who evaded disease and disability, maintained their cognitive reserve, and proactively engaged with life's pursuits [8]. Following this, the Network on an Ageing Society funded by the Mac Arthur Foundation (USA) brought researchers from diverse fields to work on the aging population by providing evidence-based solutions to the policymakers. However, their definition of successful aging has been criticized as narrow, as their research predominantly involved the top 30% of the population who have been successful their entire lives in social, educational, financial, and occupational spheres [6].

An alternate multi-criteria model posits that successful aging is a highly personal, subjective process rather than a normative outcome where the elderly dynamically adapts and optimize their reserves and expectations to compensate for the losses of aging, referred to as "selective optimization with compensation (SOC)" [9]. On the other hand, socio-centric models of successful aging have underscored the importance of inter-generational cohesion, close-knit communities, and community perception of the aged [10]. These individual and social models of aging are complimentary and impart a holistic view of the individual as well as the societal response to aging.

3.3 Economic Costs of Geriatric Healthcare Programs

Health economics has increasingly been used to evaluate the feasibility and outcome of health interventions as even high-income nations are faced with budgetary constraints and accountability to the public. Since healthcare resources are premium, careful economic analysis is paramount, especially for psychosocial interventions, which are usually resource intense, to aid policymakers and managers to allocate available resources [11].

The models of health economic evaluations can be considered broadly under three headings viz., the money spent—to the health outcomes achieved (cost-benefit analysis); the money spent—to the healthy life added in years (cost-utility analysis); and finally cost-benefit analysis, a ratio of the monetary benefit of treatment to the monetary costs of treatment [12].

3.4 Depression in the Elderly

Late-life depression is a common mental health issue that negatively impacts all the domains of quality of life of the elderly and is a significant contributor to early mortality. A recent meta-analysis estimated the global prevalence of clinical depression to be 13.3% (95% CI: 8.4–20.3%) [13]. Qualitatively, the lived experience of late-life depression is different from that of adults or the young population. The depressed elderly present with more bodily symptoms especially pain, fatigue, poor appetite, subjective memory loss, hopelessness, loss of will to continue living, and worthlessness [14]. Subsyndromal depression is five times more common in the elderly and is frequently missed in clinical assessments that use standardized tools meant for screening for adult depression and diagnostic criteria [15].

Psychosocial interventions for addressing late-life depression can be classified in many ways. In this chapter, since feasibility and cost-effectiveness are pivotal, those interventions that are cost-effective and implementable in resource-poor settings are given priority.

3.5 Anxiety Disorders in Old Age

Geriatric anxiety presents unique challenges due to the interaction between age-related changes, physical health conditions, and psychosocial factors [16]. Anxiety in old age is experienced as a spectrum ranging from appropriate worries about senescence to disabling worries and avoidance behaviors that prevent engagement with life [17]. Prevalence rates vary depending on the specific anxiety disorder and the assessment methods used. Generalized Anxiety Disorder (GAD) is one of the most prevalent of anxiety disorders in the elderly population. Studies have reported prevalence rates of GAD ranging from 3% to 14% among older adults [18]. Other anxiety disorders, such as panic disorder, phobias, obsessive-compulsive disorder (OCD), and post-traumatic stress disorder (PTSD), also occur in the geriatric population, although at lower rates compared to GAD. Geriatric anxiety can have a significant impact on both physical and mental health outcomes. Older adults with anxiety are more likely to experience comorbid physical health conditions, such as cardiovascular diseases, respiratory disorders, and chronic pain [19]. Anxiety has been associated with increased healthcare utilization, higher rates of functional decline, and increased mortality in older adults [20]. Furthermore, geriatric anxiety often co-occurs with other mental health conditions, such as depression. The presence of comorbid anxiety and depression can exacerbate symptoms, impair daily functioning, and reduce the overall quality of life for older adults [21].

3.6 Cognitive Impairment and Dementia

Cognitive decline is an inevitable part of the aging process. It is a significant health concern affecting a growing number of elderly individuals worldwide. As population aging is increasing worldwide, healthcare systems are faced with burgeoning costs of caring for the elderly affected by impairment in memory, attention, language, and problem-solving skills. Cognitive impairment presents as a spectrum ranging from a mild cognitive deficit, and delirium, to full-blown dementia.

Cognitive decline leading to dementia is already becoming a global challenge in high-income countries as it imposes a substantial burden on healthcare resources. Direct costs to public healthcare are compounded by indirect socioeconomic costs that are considerable in terms of medical treatment, equipment, home safety modifications, protective services, supported personal care, daycare facilities, residential care centers, and caregiver support services [22]. A recent meta-analysis of the cost of dementia care in low middle-income countries (LMICs) termed the impact on the GDP as "catastrophic" [23].

3.7 Psychosocial Interventions in the Elderly

Psychosocial interventions play a vital role in promoting the mental health and well-being of older adults [24]. These interventions focus on addressing the psychosocial factors that contribute to distress and focus on enhancing the overall quality of life in the elderly population. They can range from low-intensity interventions like psychoeducation to resource-intense individualized therapy sessions.

The delivery of psychosocial interventions to the elderly poses unique challenges [25]. Acceptability of psychosocial interventions varies based on cultural understanding of the disease state. For example, late-life depression can be perceived by the elderly as a medical problem (loss of appetite, somatic pain symptoms, loss of memory) rather than a psychological issue which can negatively impact acceptability and participation unless psychoeducation on the biopsychosocial causes of depression are inbuilt in the intervention [26]. Also, the indirect costs must be considered especially when the intervention requires the elderly to travel (or accompanied) to the intervention site. This can be remedied with digital psychosocial interventions which have gained sufficient evidence in the past decade, especially in LMICs [27].

In the setting of public healthcare, regular visits to the primary health provider for treatment of chronic medical illnesses is an ideal time for screening elderly patients for psychosocial problems. Care delivery can vary from face-to-face sessions, group therapy, or internet-based remote sessions.

Research on community-based psychosocial interventions for the elderly in high-income countries has clearly established its importance. However, in LMICs, data on the efficacy, cost-effectiveness, and implementation are ongoing but encouraging [28]. The following section will briefly allude to the various evidence-based psychosocial interventions and their applications in the geriatric population.

3.7.1 Cognitive Behavior Therapy

Cognitive Behavioral Therapy (CBT) is one of the most researched psychosocial interventions that address emotional and behavioral problems in mental health. It is based on the cognitive model of psychology which posits that our emotional response is influenced by our habitual thinking and behavioral patterns. By focusing on and making clients aware of the connection between thoughts, emotions, and behaviors, CBT aims to replace maladaptive beliefs, negative thoughts, and self-defeating behaviors with more realistic and rational ones. Techniques such as cognitive restructuring, goal setting, rewarding task completion, problem solving are part of the CBT process. By applying these insights in real-world situations and the feedback received, clients optimize, recalibrate, and improve on them while maintaining a diary of the thoughts and emotions that they experience while doing so helping them to track their progress [26–28].

By focusing on the present, CBT addresses thoughts, emotions, and behaviors that are maladaptive, an approach that is suited for the elderly, as it allows them to focus on their challenges and find practical solutions to enhance their mental well-being [29]. As CBT typically follows a structured and time-limited format, it is well-suited for older adults who may prefer a more focused and goal-oriented approach [30]. The limited duration of therapy provides a sense of structure and allows for targeted interventions to address specific concerns efficiently. By emphasizing on practical skills that can be applied directly to everyday life, CBT is particularly relevant for the elderly, as it promotes independence and empowers older adults to actively participate in their mental healthcare.

CBT is highly adaptable, flexible, and can be proven to address a range of concerns relevant to the elderly including depression, cognitive impairment, loneliness, insomnia, chronic pain, disability, grief, and loss, to promote physical and psychological well-being, and prevent geriatric depression and anxiety [31–34].

More importantly, CBT has been shown to improve outcomes when combined with medications, delivered remotely via telehealth platforms, in group settings by trained paramedical teams in resource-poor settings [35–37].

3.7.2 Interpersonal Therapy

Interpersonal Therapy (IPT) is a cost-effective, brief psychological intervention to address psychosocial distress by focusing on the impact of interpersonal relationships on mental health and well-being [38]. IPT operates on the premise that the quality of our relationships profoundly influences our emotional and psychological well-being. By targeting specific interpersonal issues, IPT aims to alleviate distress, enhance social functioning, and foster resilience [38].

IPT offers tailored support for older adults, facilitating personal growth, resilience, and improved QoL. Aging is inevitably accompanied by significant transitions, such as retirement, loss of loved ones, declining health, and changing roles within families. These factors can lead to feelings of isolation, less valued, loneliness, and reduced social engagement [39]. IPT assists the elderly in adapting to these life changes, fostering resilience, and helping them explore new roles and possibilities [40].

IPT sessions can be delivered remotely via the Internet, thereby accessing remote areas and the elderly who are unable to attend face-to-face sessions. Tele-IPT thus has the advantage of increased adherence and retention, comparable efficacy, greater satisfaction, and cost-effectiveness [41]. Group IPT sessions have proven to be promising in improving mental health outcomes in LMICs where manpower resources are limited [42]. Finally, IPT can be adapted to suit the elderly with cognitive impairment (IPT-ci) and address the burden among the caregivers of the elderly [43].

3.7.3 Acceptance and Commitment Therapy

Acceptance and Commitment Therapy (ACT) is an evidence-based psychological intervention to reduce distress that focuses on improving psychological flexibility [44]. Being psychologically inflexible is characterized by living in the past or future, avoidance of unpleasant thoughts and feelings, attachment to past values that are no more relevant, losing focus on values that give meaning to life, and inaction. ACT is suited for older adults as they accumulate losses in functioning and need to be psychologically flexible and resilient to successful ageing as conceptualized by the selective optimization with compensation (SOC) model [45]. The focus of ACT is to help older adults to accommodate these permanent changes that are a part of aging without avoidance or unrealistic expectations and demands and replacing them with feasible goals that add value to life.

ACT interventions may be more suited for the elderly who have lower awareness of psychological problems and mental health issues [46]. While CBT interventions are effective in adults it requires considerable psychological sophistication and awareness in recognizing, challenging, relabeling, and substituting maladaptive internal thoughts and feelings. The goal of an ACT intervention is to live life in accordance with deeply held values and this may resonate more with older adults [47]. This in turn may potentially make it more likely they will actively engage in treatment with lower attrition.

ACT has also been found effective in addressing a plethora of psychosocial challenges unique to old age including depression, geriatric anxiety, death anxiety, chronic illness and pain, disability, impairment in activities of daily living, poor QoL, and in addressing the well-being of caregivers of older adults [48–52].

ACT is also suited for delivering care in a group-based setting which is more cost-effective, time-efficient, and adaptable to LMICs, with the added benefit of addressing loneliness and isolation common in the elderly and providing an opportunity to learn from others' experiences and modeling new coping strategies and behaviors. Moreover, ACT is effective over digital and internet platforms thereby addressing rural and remote communities [53]. Integration of ACT in busy general medical practice and primary healthcare settings has been possible through brief modified ACT intervention called Focused ACT (fACT) where typically sessions last 20 min with 4–6 sessions [54].

3.8 Social Support Services for the Elderly

Social isolation and loneliness are common lived experiences in the elderly contributing to a poor QoL [55]. Psychosocial interventions in the form of senior citizen clubs and activities have been shown to reduce loneliness but are mostly deficient in LMICs. However, even relatively low-intensity interventions like providing technology-friendly mobile phones with call and internet credit are beneficial in improving the well-being of the elderly [55].

Restoring dignity and a sense of pride in the elderly is a key determinant of QoL. As an active adult who was the breadwinner, respected by the family, and in charge of decisions experiences the loss of his/her importance. Psychosocial interventions that provide opportunities for elders to sustain or rebuild their ability to contribute to their communities have been shown to restore purpose, dignity, and psychological health [56]. In LMICs innovative psychosocial programs that involve childcare or work opportunities, even if the productivity is low can have a significant impact on the QoL of the elderly [57].

3.9 Psychosocial Interventions to Improve the Physical Health and Disability

It is undeniable that physical and psychological mental health share a robust bidirectional relationship. Chronic pain is a universal experience among the aged population and is a strong predictor of poor QoL [58]. Addressing chronic pain by treating the underlying physical condition combined with psychosocial interventions that focus on coping has been shown to yield optimal outcomes [59]. In resource-high settings psychosocial interventions like CBT have shown promise [60]. In resource-poor settings, a multi-disciplinary approach incorporating multidimensional public health measures including psychoeducation about the biopsychosocial model of disease, lifestyle modifications, graded physical activity integrated with medical management have been shown to improve outcomes [61].

Aging inevitably is associated with the loss of sensory abilities. Declining vision, hearing, and mobility act as barriers preventing the elderly to socialize and communicate. These in turn create a vicious cycle of isolation and loneliness [62]. In low-income settings, where access to healthcare is limited, the elderly are often unaware that they have functional limitations or that they can be treated. Public health policies that provide for active screening, and provision of affordable and quality hearing aids, glasses, and walking aids have been shown to increase productivity, independence, and well-being [63]. Harnessing digital technology to support ongoing medical care can be implemented on a community level with low input costs. Mobile phone applications that are designed to track disease progress, remote monitoring, and medication reminders have been shown to improve medical outcomes [64].

Physical activity in the form of group exercise programs in the elderly has been shown to improve not only their physical health but overall well-being, and QoL [65]. By providing opportunities for interaction and engagement, group activities are ideal for combating loneliness and isolation in the elderly. Additionally, group exercise was found to have positive effects on cognitive function and reduced risk of cognitive decline in older adults [66].

3.10 Programs that Focus on Elder Abuse

Elder abuse refers to either intentional or unintentional detrimental acts toward the elderly by a caregiver. These acts include physical, psychological, sexual abuse, financial exploitation, and neglect [67]. It is a global problem as 1 in 6 community-dwelling older adults experienced abuse in some form over the past year [68]. A meta-analysis of psychosocial interventions targeting elder abuse has reported encouraging outcomes [69]. Multilayered interventions involving the elders, family, and caregivers with multimodal interventions including building awareness, support services, addressing social, health, financial, and legal needs, caregiver stress, and psychological support have yielded tangible benefits.

3.11 Social Programs Addressing Financial Security

Direct monetary transfers and senior citizens' cash pensions have been found to prevent or improve psychological health in both high and LMICs [70]. Regular non-contributory old-age pensions offer a promising public health policy that has been shown to reduce symptoms of depression in many LMICs [71].

3.12 Indigenous Solutions in Resource-Poor Settings

As discussed earlier, a large body of evidence has shown that specialized psychosocial interventions like CBT, IPT, and ACT effectively reduce depression and anxiety among the elderly. However, in remote and rural parts of LMICs plagued by a shortage of trained mental health professionals (psychiatrists, psychiatric nurses, clinical psychologists, and psychiatric social workers) innovative solutions are required.

To fill this void, a collaborative stepped-care approach has been suggested. In this model, relatively basic interventions such as psychoeducation and supportive therapy are provided to all elderly, while more resource-intensive interventions are reserved for moderately to severely ill screened by lay health counselors or those that do not respond to basic interventions. The collaborative approach involves a three-layer team consisting of lay health counselors, a primary care physician, and a visiting psychiatrist [72]. Such programs have been successfully implemented in Goa, India in patients with elderly dementia [73].

Another promising approach is to train the family and neighbors in delivering simple therapy booster sessions, or by integrating therapy and boosters into regular healthcare or social welfare checks. This has been integrated into the public health delivery systems in Southern India (Tamil Nadu), where village health volunteers visit eligible homes to conduct health tests, deliver medication, and supportive psychosocial interventions including psychoeducation and psychotherapy [55]. Similar interventions have also been found to prevent mental health problems like depression ([74]).

3.13 Conclusion

> Our society must make it right and possible for old people not to fear the young or be deserted by them, for the test of a civilization is the way that it cares for its helpless members.—Pearl S. Buck

This chapter illustrates the evidence base for the effectiveness of various psychosocial interventions aimed to enhance the QoL of the elderly population. These interventions encompass a wide range of approaches such as structured brief psychotherapies, counseling, social support groups, social engagement services, physical health and mobility, and financial and occupation support programs that can be modified to suit the cultural context of rural populations and LMICs. By providing emotional support, enhancing social connections, and fostering a sense of purpose and belonging, these interventions contribute to reducing loneliness, depression, and cognitive decline in the elderly to empower them, their families, and caregivers to navigate the challenges of aging and promote successful aging. Implementing and expanding access to psychosocial interventions for the elderly should be a priority in healthcare and social service systems to ensure optimal mental and emotional well-being in this vulnerable population.

Competing Interest The authors have no conflicts of interest to declare that are relevant to the content of this chapter.

References

1. Ince YM. Economic and social consequences of population aging the dilemmas and opportunities in the twenty-first century. Appl Res Qual Life. 2015;10:735–52.
2. Timonen V. Ageing societies: a comparative introduction. McGraw-Hill Education (UK); 2008.
3. Sheiner L. The determinants of the macroeconomic implications of aging. Am Econ Rev. 2014;104(5):218–23.
4. Bülow MH, Söderqvist T. Successful ageing: a historical overview and critical analysis of a successful concept. J Aging Stud. 2014;31:139–49.
5. Bowling A, Dieppe P. What is successful ageing and who should define it? BMJ. 2005;331(7531):1548–51.
6. Lupien SJ, Wan N. Successful ageing: from cell to self. Philos Trans R Soc Lond B Biol Sci. 2004;359(1449):1413–26.
7. Rowe JW, Kahn RL. Human aging: usual and successful. Science. 1987;237(4811):143–9.
8. Rowe JW, Kahn RL. Successful aging 2.0: conceptual expansions for the 21st century. J Gerontol B. 2015;70(4):593–6.
9. Baltes PB, Baltes MM. Psychological perspectives on successful aging: The model of selective optimization with compensation. In: Successful aging: Perspectives from the behavioral sciences. Cambridge: Cambridge University Press; 1990. p. 1–34.

10. Leviatan U. Contribution of social arrangements to the attainment of successful aging—the experience of the Israeli kibbutz. J Gerontol Ser B Psychol Sci Soc Sci. 1999;54(4):P205-13.
11. Knapp M, Wong G. Economics and mental health: the current scenario. World Psychiatry. 2020;19:3–14.
12. Drummond MF, Sculpher MJ, Claxton K, Stoddart GL, Torrance GW. Methods for the economic evaluation of health care programmes. Oxford university press; 2015.
13. Abdoli N, Salari N, Darvishi N, Jafarpour S, Solaymani M, Mohammadi M, Shohaimi S. The global prevalence of major depressive disorder (MDD) among the elderly: a systematic review and meta-analysis. Neurosci Biobehav Rev. 2022;132:1067.
14. Zenebe Y, Akele B, Necho M. Prevalence and determinants of depression among old age: a systematic review and meta-analysis. Ann General Psychiatry. 2021;20(1):1–9.
15. Oh DJ, Han JW, Kim TH, Kwak KP, Kim BJ, Kim SG, Kim JL, Moon SW, Park JH, Ryu SH, Youn JC. Epidemiological characteristics of subsyndromal depression in late life. Aust N Z J Psychiatry. 2020;54(2):150–8.
16. Beekman AT, de Beurs E, van Balkom AJ, Deeg DJ, van Dyck R, van Tilburg W. Anxiety and depression in later life: co-occurrence and communality of risk factors. Am J Psychiatry. 2000;157(1):89–95.
17. Grenier S, Préville M, Boyer R, O'Connor K, Béland SG, Potvin O. The impact of DSM-IV symptom and clinical significance criteria on the prevalence estimates of subthreshold and threshold anxiety in the older adult population. Am J Geriatr Psychiatry. 2010;18(6):483–96.
18. Havens CM, Hall MJ. Anxiety disorders in older adults: a comprehensive review. Geriatric Nurs. 2019;40(3):254–61.
19. El-Gabalawy R, Mackenzie CS, Shooshtari S, Sareen J. Comorbid physical health conditions and anxiety disorders: a population-based exploration of prevalence and health outcomes among older adults. Gen Hosp Psychiatry. 2011;33(6):556–64.
20. Wolitzky-Taylor KB, Castriotta N, Lenze EJ, Stanley MA, Craske MG. Anxiety disorders in older adults: a comprehensive review. Depress Anxiety. 2010;27(2):190–211.
21. Byers AL, Yaffe K, Covinsky KE, Friedman MB, Bruce ML. High occurrence of mood and anxiety disorders among older adults: the National Comorbidity Survey Replication. Arch Gen Psychiatry. 2010;67(5):489–96.
22. Park HL, O'Connell JE, Thomson RG. A systematic review of cognitive decline in the general elderly population. Int J Geriatr Psychiatry. 2003;18(12):1121–34.
23. Mattap SM, Mohan D, McGrattan AM, Allotey P, Stephan BC, Reidpath DD, Siervo M, Robinson L, Chaiyakunapruk N. The economic burden of dementia in low-and middle-income countries (LMICs): a systematic review. BMJ Glob Health. 2022;7(4):e007409.
24. Sutipan P, Intarakamhang U, Macaskill A. The impact of positive psychological interventions on well-being in healthy elderly people. J Happiness Stud. 2017;18:269–91.
25. Lovell K, Lamb J, Gask L, Bower P, Waheed W, Chew-Graham C, Lamb J, Aseem S, Beatty S, Burroughs H, Clarke P. Development and evaluation of culturally sensitive psychosocial interventions for under-served people in primary care. BMC Psychiatry. 2014;14:1–6.
26. Barbui C, Purgato M, Abdulmalik J, Acarturk C, Eaton J, Gastaldon C, Gureje O, Hanlon C, Jordans M, Lund C, Nose M. Efficacy of psychosocial interventions for mental health outcomes in low-income and middle-income countries: an umbrella review. Lancet Psychiatry. 2020;7(2): 162–72.
27. Scazufca M, Nakamura CA, Peters TJ, Henrique MG, Seabra A, La Rotta EG, Franzin RM, Martins DF, Van de Ven P, Hollingworth W, Araya R. A collaborative care psychosocial intervention to improve late life depression in socioeconomically deprived areas of Guarulhos, Brazil: the PROACTIVE cluster randomised controlled trial protocol. Trials. 2020;21(1):1–4.
28. Giebel C, Shrestha N, Reilly S, White RG, Zuluaga MI, Saldarriaga G, Liu G, Allen D, Gabbay M. Community-based mental health and well-being interventions for older adults in low-and middle-income countries: a systematic review and meta-analysis. BMC Geriatr. 2022;22(1): 773.

29. Jayasekara R, Procter N, Harrison J, Skelton K, Hampel S, Draper R, Deuter K. Cognitive behavioural therapy for older adults with depression: a review. J Ment Health. 2015;24(3): 168–71.
30. Simon SS, Cordás TA, Bottino CM. Cognitive behavioral therapies in older adults with depression and cognitive deficits: a systematic review. Int J Geriatr Psychiatry. 2015;30(3): 223–33.
31. Jarvis MA, Padmanabhanunni A, Chipps J. An evaluation of a low-intensity cognitive behavioral therapy mHealth-supported intervention to reduce loneliness in older people. Int J Environ Res Public Health. 2019;16(7):1305.
32. McLaren DM, Evans J, Baylan S, Smith S, Gardani M. The effectiveness of the behavioural components of cognitive behavioural therapy for insomnia in older adults: a systematic review. J Sleep Res. 2023;19:e13843.
33. Kerns RD, Otis JD, Marcus KS. Cognitive-behavioral therapy for chronic pain in the elderly. Clin Geriatr Med. 2001;17(3):503–23.
34. Hall J, Kellett S, Berrios R, Bains MK, Scott S. Efficacy of cognitive behavioral therapy for generalized anxiety disorder in older adults: systematic review, meta-analysis, and meta-regression. Am J Geriatr Psychiatry. 2016;24(11):1063–73.
35. Wetherell JL, Petkus AJ, White KS, Nguyen H, Kornblith S, Andreescu C, Zisook S, Lenze EJ. Antidepressant medication augmented with cognitive-behavioral therapy for generalized anxiety disorder in older adults. Am J Psychiatr. 2013;170(7):782–9.
36. Andersson G, Topooco N, Havik O, Nordgreen T. Internet-supported versus face-to-face cognitive behavior therapy for depression. Expert Rev Neurother. 2016;16(1):55–60.
37. Wuthrich VM, Rapee RM. Randomised controlled trial of group cognitive behavioural therapy for comorbid anxiety and depression in older adults. Behav Res Ther. 2013;51(12):779–86.
38. Bosmans JE, Van Schaik DJ, Heymans MW, Van Marwijk HW, Van Hout HP, De Bruijne MC. Cost-effectiveness of interpersonal psychotherapy for elderly primary care patients with major depression. Int J Technol Assess Health Care. 2007;23(4):480–7.
39. Miller MD, Cornes C, Frank E, Ehrenpreis L, Silberman R, Schlernitzauer MA, Tracey B, Richards V, Wolfson L, Zaltman J, Bensasi S. Interpersonal psychotherapy for late-life depression: past, present, and future. J Psychother Pract Res. 2001;10(4):231.
40. Miller MD. Using interpersonal therapy (IPT) with older adults today and tomorrow: a review of the literature and new developments. Curr Psychiatry Rep. 2008;10(1):16–22.
41. Donker T, Bennett K, Bennett A, Mackinnon A, van Straten A, Cuijpers P, Christensen H, Griffiths KM. Internet-delivered interpersonal psychotherapy versus internet-delivered cognitive behavioral therapy for adults with depressive symptoms: randomized controlled noninferiority trial. J Med Internet Res. 2013;15(5):e82.
42. Verdeli H, Clougherty K, Bolton P, Speelman L, Lincoln N, Bass J, Neugebauer R, Weissman MM. Adapting group interpersonal psychotherapy for a developing country: experience in rural Uganda. World Psychiatry. 2003;2(2):114.
43. Miller MD, Reynolds CF. Expanding the usefulness of interpersonal psychotherapy (IPT) for depressed elders with co-morbid cognitive impairment. Int J Geriatr Psychiatry. 2007;22:101–5.
44. Ciarrochi J, Bilich L, Godsell C. Psychological flexibility as a mechanism of change in acceptance and commitment therapy. In: Baer RA (Ed). Assessing mindfulness and acceptance processes in clients: Illuminating the theory and practice of change. Oakland, CA: Context Press; 2010. p 51–75.
45. Alonso-Fernández M, López-López A, Losada A, González JL, Wetherell JL. Acceptance and commitment therapy and selective optimization with compensation for institutionalized older people with chronic pain. Pain Med. 2016;17(2):264–77.
46. Hyer L, Rafalson L, O'Hea EL. Selected optimization with compensation: older adults adjusting to change. In Handbook of stress, trauma, and the family 2013 Jan 11 (pp. 69-88). Routledge.
47. Petkus AJ, Wetherell JL. Acceptance and commitment therapy with older adults: rationale and considerations. Cogn Behav Pract. 2013;20(1):47–56.

48. Twohig MP, Levin ME. Acceptance and commitment therapy as a treatment for anxiety and depression: a review. Psychiatr Clin. 2017;40(4):751–70.
49. Ahmadi V, Valizadeh H. The effectiveness of acceptance and commitment-based therapy on the quality of life and death anxiety in the elderly. Aging Psychol. 2021;7(2):166–53.
50. Hashemi Z, Afshari A, Einy S. The effectiveness of acceptance and commitment education on improving the mental health and quality of life of elderly people with cancer. Iran J Health Educ Health Promot. 2020;8(2):160–71.
51. Lai L, Liu Y, McCracken LM, Li Y, Ren Z. The efficacy of acceptance and commitment therapy for chronic pain: a three-level meta-analysis and a trial sequential analysis of randomized controlled trials. Behav Res Ther. 2023;165:104308.
52. Lappalainen P, Pakkala I, Lappalainen R, Nikander R. Supported web-based acceptance and commitment therapy for older family caregivers (CareACT) compared to usual care. Clin Gerontol. 2022;45(4):939–55.
53. Reijnders JS, Batink T, Simons ME, Peeters SC, Janssens M, Lataster JJ, Jacobs NE. A preventive eHealth ACT module for positive aging: feasibility results. In Association for Contextual Behavioral Science Virtual World Conference 2021 2021 Jun 25.
54. Glover NG, Sylvers PD, Shearer EM, Kane MC, Clasen PC, Epler AJ, Plumb-Vilardaga JC, Bonow JT, Jakupcak M. The efficacy of focused acceptance and commitment therapy in VA primary care. Psychol Serv. 2016;13(2):156.
55. Banerjee A, Duflo E, Grela E, McKelway M, Schilbach F, Sharma G, Vaidyanathan G. Depression and loneliness among the elderly in LMICs. J Econ Perspect. 2023;37(2): 179–202.
56. Hussam R, Kelley EM, Lane G, Zahra F. The psychosocial value of employment: evidence from a refugee camp. Am Econ Rev. 2022;112(11):3694–724.
57. Shofoyeke AD, Amosun PA. A survey of care and support for the elderly people in Nigria. Mediterr J Soc Sci. 2014;5(23):2553.
58. Ferretti F, Castanha AC, Padoan ER, Lutinski J, Silva MR. Quality of life in the elderly with and without chronic pain. BrJP. 2018;1:111–5.
59. Miaskowski C, Blyth F, Nicosia F, Haan M, Keefe F, Smith A, Ritchie C. A biopsychosocial model of chronic pain for older adults. Pain Med. 2020;21(9):1793–805.
60. Lunde LH, Nordhus IH, Pallesen S. The effectiveness of cognitive and behavioural treatment of chronic pain in the elderly: a quantitative review. J Clin Psychol Med Settings. 2009;16:254–62.
61. Siddiqui AS, Bashir U. Teamwork in chronic pain management and the way forward in low and middle-income countries. Anaesth Pain Intens Care. 2021;25(2):229–35.
62. Marmamula S, Kumbham TR, Modepalli SB, Barrenkala NR, Yellapragada R, Shidhaye R. Depression, combined visual and hearing impairment (dual sensory impairment): a hidden multi-morbidity among the elderly in residential Care in India. Sci Rep. 2021;11(1):1–9.
63. Ye X, Zhu D, Chen S, Shi X, Gong R, Wang J, Zuo H, He P. Effects of providing free hearing aids on multiple health outcomes among middle-aged and older adults with hearing loss in rural China: a randomized controlled trial. BMC Med. 2022;20(1):1–2.
64. Choi A, Lovett AW, Kang J, Lee K, Choi L. Mobile applications to improve medication adherence: existing apps, quality of life and future directions. Adv Pharmacol Pharm. 2015;3 (3):64–74.
65. Rebar AL, Stanton R, Geard D, Short C, Duncan MJ, Vandelanotte C. A meta-meta-analysis of the effect of physical activity on depression and anxiety in non-clinical adult populations. Health Psychol Rev. 2015;9(3):366–78.
66. Physical activity and enhanced fitness to improve cognitive function in older people without known cognitive impairment. Cochrane Database Syst Rev. 2008;2:CD005381.
67. Pavlik VN, Hyman DJ, Festa NA, Dyer CB. Quantifying the problem of abuse and neglect in adults—analysis of a statewide database. J Am Geriatr Soc. 2001;49(1):45–8.
68. Yon Y, Mikton CR, Gassoumis ZD, Wilber KH. Elder abuse prevalence in community settings: a systematic review and meta-analysis. Lancet Glob Health. 2017;5(2):e147–56.

69. Shen Y, Sun F, Zhang A, Wang K. The effectiveness of psychosocial interventions for elder abuse in community settings: a systematic review and meta-analysis. Front Psychol. 2021;12: 679541.
70. Ridley M, Rao G, Schilbach F, Patel V. Poverty, depression, and anxiety: causal evidence and mechanisms. Science. 2020;370(6522):eaay0214.
71. Bando R, Galiani S, Gertler P. The effects of noncontributory pensions on material and subjective Well-being. Econ Dev Cult Chang. 2020;68(4):1233–55.
72. Patel V, Weiss HA, Chowdhary N, Naik S, Pednekar S, Chatterjee S, De Silva MJ, Bhat B, Araya R, King M, Simon G. Effectiveness of an intervention led by lay health counselors for depressive and anxiety disorders in primary care in Goa, India (MANAS): a cluster randomized controlled trial. Lancet. 2010;376(9758):2086–95.
73. Dias A, Dewey ME, D'Souza J, Dhume R, Motghare DD, Shaji KS, Menon R, Prince M, Patel V. The effectiveness of a home care program for supporting caregivers of persons with dementia in developing countries: a randomized controlled trial from Goa, India. PLoS One. 2008;3(6): e2333.
74. Barker N, Bryan G, Karlan D, Ofori-Atta A, Udry C. Cognitive Behavioral therapy among Ghana's rural poor is effective regardless of baseline mental distress. Am Econ Rev Insights. 2022;4(4):527–45.

Chapter 4
The Economic Impact of Ageing on Healthcare

Polly Lama

Abstract The economic value of longevity and the rise in the ageing population are directly related, and is approximately estimated to incur 9.1 trillion dollars globally as expenses over the next 50 years. Total government operating costs, including social security and pension pay will therefore rise substantially with delayed ageing. As presented here, healthcare and pension policies expenditure varies between high, middle, and low-income countries, and the current system indicate shortages to meet the rising cost of care and living as the ageing group inflates, these can be offset modestly by implementing newer policies, which may include plans to reserve national savings by adopting strategies such as expanded investment schemes apart from the pension plan for self-employed, non-employed, and the retired individuals. Integration of healthcare, social care programmes, and long-term care facilities at an early stage is another prospect likely to perhaps be a efficient way to forestall disease and extend economic prosperity towards leading a healthy and financially viable retired life.

Keywords Ageing · Expenditure · Expenses · Income · Pension

4.1 Introduction

Planning productive ageing policies is the main 'rationale' behind progressive development for the aged, retired, and the oldest old, further, 2021–2030 has been declared by the United Nations as the 'Decade of Healthy Aging' [1], a global movement aimed to connect different stakeholders towards the common goal of providing a better quality of life, healthcare facilities and socioeconomic opportunities for the elderly.

The continent-wise ageing patterns, reform measures adopted, and healthcare structure are as described in Chap. 1, and the advocated policies confirm variations,

P. Lama (✉)
Sikkim Manipal Institute of Medical Sciences, Sikkim Manipal University, Gangtok, Sikkim, India
e-mail: pollylama@smims.smu.edu.in

P. Lama (ed.), *The Ageing Population*,
https://doi.org/10.1007/978-981-99-5772-9_4

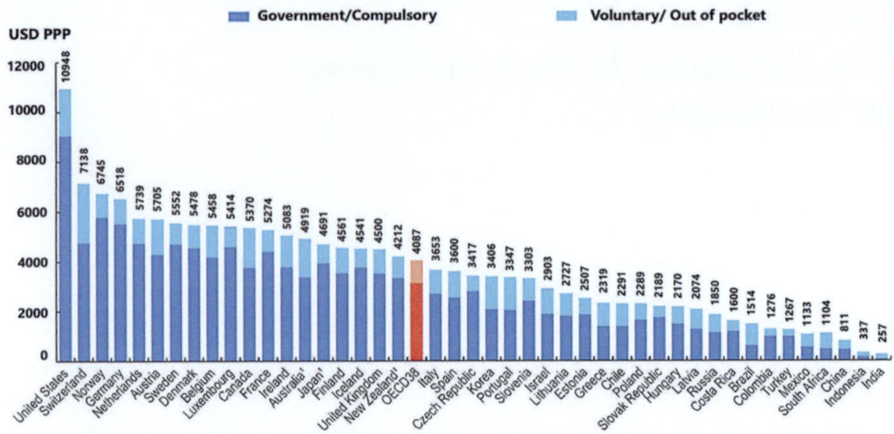

Fig. 4.1 Heath expenditure per capita in low-, middle-, and high-income economies, 2019 (or nearest years). OECD Health Statistics 2021, WHO Global Health Expenditure Database [5]

especially in terms of healthcare facilities and services between developing and the developed countries (Fig. 4.1), and their continents. Published data on increasing ageing population and economic prosperity indicates that the retired and the oldest old group may act as a source of pressure, especially on the healthcare sector [2, 3], and can influence economic viability if ageing reforms are not adequality planned [4], as the increasing proportions of ageing group influence epidemiological susceptibility to diseases, disabilities, and age-acquired deformities [5, 6]. The significant impact of healthcare system and ageing was evidently experienced during the COVID-19 outbreak, as the aged group was twice as likely to acquire infections following exposure and were more susceptible to death after exposure [7]. Perhaps, one of the most important changes linked with age is weakening of the innate immune system and vulnerability towards acquiring communicable and non-communicable diseases and disabilities [8, 9], which is adverse in countries with low economic profiles. Dependence on long-term care, especially amongst the oldest old group who are unable to work or sustain themselves additionally burdens the healthcare support staff tasks and costs [10, 11]. Thus, defining the prospect of 'healthy ageing' varies between continents and economic status of the countries, and these variabilities present challenges towards developing a universally acceptable solution for the aged healthcare system. Many advanced high-income countries have robust healthcare policies with centre for ageing or geriatric care [12, 13], whilst one of the common features of moderate, and low-income economies is the struggle to develop proper primary hospital care facilities, especially in remote rural locations of various developing countries and continents [14, 15]. Thus, the aim of this chapter is to describe the challenges, opportunities, and differences in the ageing economy with relevance to expenditure incurred by the healthcare system in low-, middle-, and high-income countries to suggest the most cost-effective healthcare practices.

Enabling the elderly to live independently and productively is essential for the economic development of any country [16, 17], and as the World population increases each year the number of older aged working, non-working group, and the frail oldest old will also increase [1, 18], therefore, factors associated with providing better healthcare facilities suitable to the needs of the ageing population is unavoidable [4, 19], and adopting a cost-effective measures may be an essential prerequisite towards sustaining the rising expenditure.

4.2 Ageing and Healthcare System in High-Income Countries

Japan [20], France [21], the United Kingdom [22, 23], the United States [24], Germany [25], South Korea [26], South Africa [27], Denmark [28], Australia [29], New Zealand [30, 31], and Netherlands [32] are some of the few countries with largest share of GDPs spending for aged care implemented policies, social facilities, and proactive reform measures [33–35].

A common and effective feature of the aged and the retired in high-income countries is they are presented with many opportunities to make the ageing individuals self-reliant and self-sufficient even after the statutory retirement ages [12, 36]. Societal prosperity is directly linked with coverage of the old-age social security system [1, 37], and earning opportunities post-retirement allows other institutional sources of financial support to broaden. Increasing employability also maintains a balance between active workers' and retiree's contribution to national savings [3]. The raised age provides incentives to work longer and to maximize pension benefits [24, 38]. In the United kingdom, ageing is correlated with general well-being and the implemented policies aim to minimize the use of public funds on pension spending [39], and working-age adults including self-employed individuals are encouraged early on to invest in privately managed pension funds with tax-free after retirement returns [40]. This adopted strategy is suggested to accumulate general state pension funds and boosts reserves for the national saving plan.

Other common strategies to promote cost-effective health and long-term care include the use of integrated and automated technology for screening the health of the still employable elderly on a regular basis, and long-term care offered to the oldest old are enabled with flexible homecare-based model allowing better treatment and care choices for the frail nearing towards the end of life [41].

A cost-effective healthcare system and disease prevention throughout life and at older ages are also achieved by targeting behavioural risk factors and administration of appropriate therapeutic drugs at an appropriate older age (Fig. 4.2). In developed countries, this is achieved through active healthcare checks, social welfare plans, and public awareness programmes. For example, in the United States, declines in disability at older ages between the 1980s and early 2000s were driven in part by a decline in deaths due to heart and circulatory system complications, which were

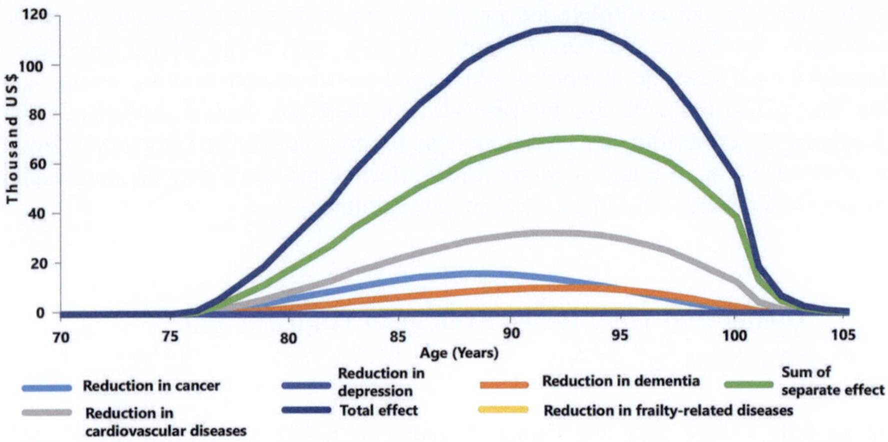

Fig. 4.2 Improvements in the incidence of various diseases under simulated impact of metformin as a prescribed type-2 diabetes drug. Sum of separate effects, total effects, are the overall value for each year of health improvements attributed to metformin. Solid lines represent five diseases as co-morbidities separately. Scott et al., The Economic Impact of targeting Ageing. Nature Aging, 2021 [45]

largely preventable by tackling associated risk factors such as high blood pressure, smoking, excess alcohol use, obesity, unhealthy diets, and sedentary lifestyles [42, 43].

Population studies on healthy lifestyle also suggest that those who quit smoking at 65 years of age live longer [44, 45] and have lower rates of pulmonary conditions, cardiovascular diseases, and mild cognitive impairment than those who continue to smoke [46]. Extensive research has also linked increased physical activity in older ages with lower hazards of disability and improved balance, mobility, cognition, and well-being [47–49]. In contrast, promoting health and healthy lifestyle in older adults is yet to be a widely adopted policy priority, especially in low, and middle-income countries [49], and there are only a few interventions tailored for older people.

4.3 Ageing and Healthcare System in Middle, and Low-Income Countries

African and Asian continents form almost 65% of the low and middle-income countries in the World [50]. India and China are the two most populous countries of the World [51], however, the total number of aged individuals in the Asian, African, and some of the less advanced countries of the European and American continents, as described in Chap. 1, have in general, lesser proportions of elderlies in comparisons to high-income developed countries. The lower percentages of elderlies are due to lack of adequate healthcare facilities (Fig. 4.3), shortage of food supplies,

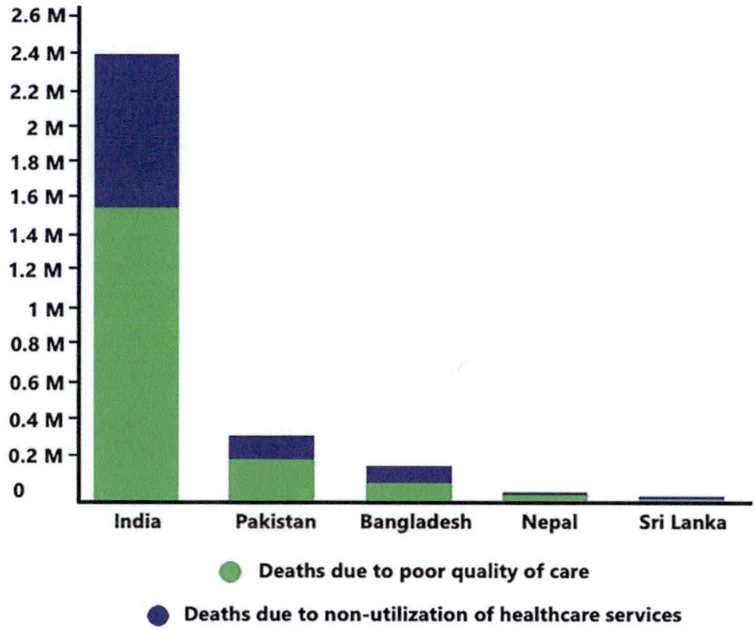

Fig. 4.3 Deaths in India and the neighbouring Asian countries due to lack of social awareness, and adequate healthcare system. Kurk et al., The Lancet (2018)

internal political conflicts, and natural calamities declining the life expectancy to an average of 75 years [16].

A common feature of the aged in the middle, and low-income economic countries, and poor health conditions are due to the greater incidence of lower literacy rates amongst the older adult [52]. Therefore, many elderlies depend on manual labour work or unpaid household services, producing outputs that have often unrecognized economic or social value [53, 54]. Older adults in less developed economies also confirm not owning a cell phone for routine communications, especially in the rural regions [9, 55]. Thus, user-friendly technology enabled healthcare operatives are beyond operational means. Older women, in the low economic countries, are mostly illiterate, and they serve their families performing domestic household functions their whole lives, without any incentives, and adequate access to routine health check-ups [56]. The lack of educational background also correlates with a greater incidence of infectious diseases such as diarrhoea, tuberculosis, pneumonia, and especially HIV/AIDS which is highly prevalent in the African and Asian continents [57, 58]. Non-communicable diseases such as diabetes, stroke, cardiovascular complications, and chronic lung diseases are other more serious and common complications on rise due to lack of hospitals, lack of proper awareness of dietary and lifestyle choices adopted by the elderlies in low, and middle-income economies.

Unlike the advanced countries, long-term aged care facilities for the oldest old who are unable to work are non-existent [34], and it is normally the families who care for them, however, essential food supplies and community-based primary medical services are provided to the oldest old with the help of the local Government's or Non-Government's organization at a rate of 92% [54], excepting in cases of severe natural calamities and war inflicted regions. These community-based medical centre often lack advanced care facilities for the overall population and the aged.

Older people, however, are not always insignificant in low, and middle-income countries with relevance to productivity and prosperity, as they do provide economic and societal benefits particularly if they are healthy and active, as they provide important inputs into caring for other older people and younger children being raised at their homes, thus relieving the care burden from the working aged adult group, and also by performing other community-related support tasks [59, 60]. Whilst undeniably oldest old nearing death brings higher needs for costs to the dependent carers, such situations can be moderated with better configurations of the current health and social care welfare programmes. Thus, two terms that offer greater precision when defining the conditions of an older adult who is not experiencing healthy ageing is 'fragility' and 'loss of intrinsic capacity' to perform independent actions.

According to WHO, health is a state of complete physical, mental, and social well-being and not merely the absence of disease or infirmity. Ageing process is associated with new and growing health challenges, and the health needs of older adults differ from those of adult individuals in many ways, population ageing must, therefore, address issues of expenditure, especially in terms of long-term care [61].

Other significant findings indicate, ageing issues are distinct and differ due to individual earning status within the same region, and country, mainly due to differences in socioeconomic status of the oldest old, which is a unique feature of many low and middle-income countries [62].

Overall, another universal common risk factor associated with ageing and expenditure, worsening the remaining life lived, is the occurrenc of musculoskeletal disorders with age [63, 64]. Osteoarthritis is considered to be concomitant with ageing across the globe. Osteoarthritis is also the third most common diagnosis amongst hospitalized older patients in the world [65], along with other serious and life-threatening diseases, such as cardiovascular complications, dementia, chronic obstructive pulmonary diseases, cancer, cerebrovascular diseases, and heart failure [66].

4.4 Healthcare Expenditure and High-Income Countries

Increase in healthcare expenditures due to ageing is suggested to depend on two factors: the difference in average spending levels between older and younger groups, and the size of the population shifting into older ages in each country and the continent [3]. The UN ageing population estimates that on an average there will be

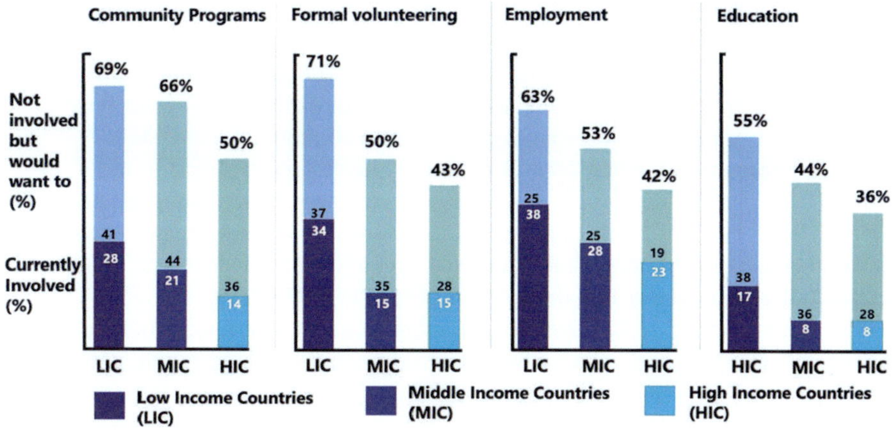

Fig. 4.4 Engagement and desire to engage in various activities amongst older adults above the age of 60 years, % of respondents ($n = 20{,}677$) [50, 71]

a 35% increase in the total number of older adults by 2050 in various high, middle, and low-income countries, however, when the cost associated with healthcare expenditure is mapped it would mean only a marginal increase of 1.5–2% of the GDP, and one of the reason behind the slow increase is, as such, ageing is far too gradual process to bring about a substantial change rapidly, and following retirement it is almost unlikely that all adults above the age of 65 years will require continuous and expensive hospital care facilities [2]. However, considering the rising total global population it would be impossible to ignore the increasing size of the population shifting into older ages [18], thus, the financial relationship between increasing age and healthcare spending can be minimized by adopting sustainable healthy lifestyle and active ageing policy, which can begin as early as from 60 years onwards. Recent evidence from the UK, the USA, Ireland, and many other European countries shows that calendar age itself is not the primary reason for the observed increases in spending correlated with age. Rather, factors, such as chronic illness, disabilities, and proximity to death, are the key determinants of higher health spending.

Other factors associated with ageing expenditure is the incentives required to train long-term care delivery experts for proper geriatric care services and the currently available manpower, thus in this regard health educational system restructuring and offering specialist courses for older people care perhaps play an important role, and cost pressures can thus arise both from improved opportunities to treat diseases common in older age and raised expectations to have more active life by the older people who are increasing gradually.

Healthcare should therefore address longevity and long-term care costs [41, 61], as the oldest old is set to increase rapidly due to better therapeutic treatment, but from a very low baseline in most countries (Fig. 4.4), and even doubling the numbers of elderlies would represents a modest absolute increase if newer cost-effective policies are to be formulated.

Organizational integration is one of the examples of a cost-effective care model, demonstrated in Torbay, England, where multidisciplinary teams from health and social care sector have been brought together into community hospitals [67]. These integrated teams were given control of pooled health services and long-term care budgets to support older people at risk of hospitalization to remain independent for longer. It is estimated that increased integration in Torbay created savings of 250,000 pounds in the first year; increased access to intermediate care; and led to a 24% fall in emergency bed day use for people aged 75 years and over. There is other evidence that case management targeting frequent emergency room attendees can be cost-effective and reduce unnecessary and expensive hospital utilization [68].

4.5 Healthcare Expenditure in Low and Middle-Income Countries

There is considerable heterogeneity within Asia, Africa, and many other Southern European, Oceanic, Northern, and Central American countries with relevance to employment after retirement, income source, coverage of Government operated old-age security pension system, education, and healthcare (Fig. 4.4). These differences are magnified, especially when considering the elderly population residing in urban and rural areas [19], and even countries at similar stages of economic prosperity, such as India and China. A key factor associated with such differences in earnings and lifestyle management in low and middle-income economies is variability in earnings acquired through one's own work, savings, inherited assets, and support offered by the family, and access to privately owned healthcare system, with advanced facilities on par with the high-income countries.

Population studies on economic viability amongst older adults in low and middle-income economics suggest that rich older adults who own large assets have substantial financial savings, and better access to privately managed care facilities, and are even supported by medical insurance schemes to cover the expenditure for major and minor illnesses, however, these privileged group forms only 30% of the older adults, whilst the remaining 70% are depended on their own earnings as manual labours, family support, Government, and non-Government organization for the healthcare expenses incurred during illness [69–71].

Dependency is especially higher amongst older women who are often uneducated and unemployed their whole lives, and they are often cared by their spouses and families. Lack of education is the primary reason for poverty and employment as manual labour in older ages. In less advanced economics, the importance of maintaining healthy lifestyles at older ages lack considerably and the implemented policies fail to bring a behavioural change towards advocating better healthy practices including maintaining proper sanitation and upgrading or implementing literacy services for older adults to combat the incidence of non-communicable diseases. Africa reports the highest number of deaths caused by diarrhoea amongst other older adults in the World. Thus, implementation of health and hygiene programmes that provides the necessary information to deal with commonly occurring and preventable illness is essential but still lack substantial efforts. Understanding of good health practices at the level of the community and the country as a whole still lacks in many such underprivileged countries.

Technology-enabled services are out of reach of many poor older adults, and even if charity programmes provide such electronic aids the uneducated elders are unable to understand how to use such devices and services. Thus, such a scenario clearly demonstrates that a lack of educational and financial ignorance carries significant costs.

4.6 Conclusion

Financial viability of health systems and the ageing population, depend on the care provided to them at some point in their lives, which will become more expensive gradually if adequate reform measures are not taken into consideration. If there is a rise in per person spending on older people relative to younger, this will create a facet in the economy and will act as a major driver of expenditure which would undermine sustainability and economic prosperity. A number of policy options are, however, available to address ageing-related healthcare expenditure. Some of these options will include greater use of innovative cost-effective technology; increased integration or coordination of care; and incentives to provide rational care towards the end of life for the frail oldest old.

Competing Interest The author has no conflicts of interest to declare that are relevant to the content of this chapter.

References

1. Amuthavalli Thiyagarajan J, Mikton C, Harwood RH, Gichu M, Gaigbe-Togbe V, Jhamba T, et al. The UN decade of healthy ageing: strengthening measurement for monitoring health and wellbeing of older people. Age Ageing. 2022;51:7.
2. Walker A, Foster L, Edward EP. The political economy of ageing and later life : critical perspectives. Cheltenham: Edward Elgar Pub. Ltd; 2014. 1 online resource (v.) p
3. United N. World population prospects 2022: summary of results. New York: United Nations; 2022.
4. Greer SL. Ageing and health: the politics of better policies. New York, NY: Cambridge University Press; 2021. pages cm. p
5. National Research Council (U.S.), Panel on a Research Agenda and New Data for an Aging World. Preparing for an aging world: the case for cross-national research. Washington, D.C.: National Academy Press; 2001. xv, 308 p. p

6. Heckles JP, Bergin LM. Aging : mental aspects, social welfare, and health. Hauppauge, N.Y: Nova Science Publishers; 2011.
7. Zhang H, Wu Y, He Y, Liu X, Liu M, Tang Y, et al. Age-related risk factors and complications of patients with COVID-19: a population-based retrospective study. Front Med (Lausanne). 2021;8:757459.
8. Hellevik S, Mustafa S, Zhang Y, Shirsat A, Saikat S. Multisectoral action towards sustainable development goal 3.d and building health systems resilience during and beyond COVID-19: findings from an INTOSAI development initiative and World Health Organization collaboration. Front. Public Health. 2023;11:1104669.
9. Skirbekk V, Dieleman JL, Stonawski M, Fejkiel K, Tyrovolas S, Chang AY. The health-adjusted dependency ratio as a new global measure of the burden of ageing: a population-based study. Lancet Healthy Longev. 2022;3(5):e332–e8.
10. Roquebert Q, Tenand M. Informal care at old age at home and in nursing homes: determinants and economic value. Eur J Health Econ. 2023;
11. Bloom DE, Chatterji S, Kowal P, Lloyd-Sherlock P, McKee M, Rechel B, et al. Macroeconomic implications of population ageing and selected policy responses. Lancet. 2015;385(9968): 649–57.
12. Papanicolas I, Woskie LR, Jha AK. Health care spending in the United States and other high-income countries. JAMA. 2018;319(10):1024–39.
13. Jivetti B, Hoque MN. SpringerLink. Population change and public policy. 1st 2020. ed. Cham: Springer International Publishing : Imprint: Springer; 2020.
14. United N. Addressing the challenges of population ageing in Asia and the Pacific: implementation of the Madrid international plan of action on ageing. New York: United Nations; 2017.
15. United Nations. Economic and Social Commission for Asia and the Pacific. Population division. Ageing and the family in the developing Asian and Pacific countries. New York, NY: United Nations; 1994. p. 73–5.
16. Christensen K, Doblhammer G, Rau R, Vaupel JW. Ageing populations: the challenges ahead. Lancet. 2009;374(9696):1196–208.
17. Collaborators GDaH. Global, regional, and national disability-adjusted life-years (DALYs) for 359 diseases and injuries and healthy life expectancy (HALE) for 195 countries and territories, 1990-2017: a systematic analysis for the Global Burden of Disease Study 2017. Lancet. 2018;392(10159):1859–922.
18. Munnell AH, Hou W, Boston College. Center for Retirement Research. Will millennials be ready for retirement? Chestnut Hill, MA: Center for Retirement Research at Boston College; 2018. Available from: http://crr.bc.edu/wp-content/uploads/2018/01/IB_18-2.pdf Available from: http://resource.nlm.nih.gov/101727768
19. World Bank (Washington District of Columbia). World Development Report 2021: data for better lives. Washington: World Bank; 2021. pages cm p
20. Arai H, Ouchi Y, Toba K, Endo T, Shimokado K, Tsubota K, et al. Japan as the front-runner of super-aged societies: perspectives from medicine and medical care in Japan. Geriatr Gerontol Int. 2015;15(6):673–87.
21. Blöss T. Ageing, lifestyles and economic crises : the new people of the Mediterranean. London ; New York: Routledge, Taylor & Francis Group; 2018. xix, 276 pages p
22. Watt T, Charlesworth A, Gershlick B. Health and care spending and its value, past, present and future. Future Healthc J. 2019;6(2):99–105.
23. Anderson GF, Davis K, Guterman S. Medicare payment reform: aligning incentives for better care. Issue Brief (Commonw Fund). 2015;20:1–12.
24. Bonoli G, Shinkawa T. Ageing and pension reform around the world : evidence from eleven countries. Cheltenham, UK; Northampton, MA: Edward Elgar; 2005. xi, 279 pp
25. Blümel M, Spranger A, Achstetter K, Maresso A, Busse R. Germany: health system review. Health Syst Transit. 2020;22(6):1–272.
26. Tan CC, Lam CSP, Matchar DB, Zee YK, Wong JEL. Singapore's health-care system: key features, challenges, and shifts. Lancet. 2021;398(10305):1091–104.

27. Younger DS. Health care in South Africa. Neurol Clin. 2016;34(4):1127–36.
28. Schmidt M, Schmidt SAJ, Adelborg K, Sundbøll J, Laugesen K, Ehrenstein V, et al. The Danish health care system and epidemiological research: from health care contacts to database records. Clin Epidemiol. 2019;11:563–91.
29. Dixit SK, Sambasivan M. A review of the Australian healthcare system: a policy perspective. SAGE Open Med. 2018;6:2050312118769211.
30. Keene L, Bagshaw P, Nicholls MG, Rosenberg B, Frampton CM, Powell I. Funding New Zealand's public healthcare system: time for an honest appraisal and public debate. N Z Med J. 2016;129(1435):10–20.
31. Nicholls MG, Frampton CM, Bagshaw PF. Resurrecting New Zealand's public healthcare system or a charity hospital in every town? Intern Med J. 2020;50(7):883–6.
32. Kroneman M, Boerma W, van den Berg M, Groenewegen P, de Jong J, van Ginneken E. Netherlands: health system review. Health Syst Transit. 2016;18(2):1–240.
33. Koohi F, Nedjat S, Yaseri M, Cheraghi Z. Quality of life among general populations of different countries in the past 10 years, with a focus on human development index: a systematic review and meta-analysis. Iran J Public Health. 2017;46(1):12–22.
34. Finkelstein EA, Bhadelia A, Goh C, Baid D, Singh R, Bhatnagar S, et al. Cross country comparison of expert assessments of the quality of death and dying 2021. J Pain Symptom Manag. 2022;63(4):e419–e29.
35. Schütte S, Acevedo PNM, Flahault A. Health systems around the world—a comparison of existing health system rankings. J Glob Health. 2018;8(1):010407.
36. Organisation for Economic Co-operation and D. Working better with age. Paris, France: OECD; 2019.
37. Organisation for Economic Co-operation and D. Pensions at a glance 2021. S.l: OECD Publishing; 2021.
38. The retirement effects of old-age pension and early retirement schemes in OECD countries. OECD Economics Department Working Papers 370, OECD Publishing. https://doi.org/10.1787/308728704511
39. Holzmann R. Aging population, pension funds, and financial markets : regional perspectives and global challenges for central, eastern, and southern Europe. Washington, DC, Vienna: World Bank; ERSTE Stiftung; 2009. xvi, 162 pp
40. Hagemann S, Scherger S. Increasing pension age - inevitable or unfeasible? Analysing the ideas underlying experts' arguments in the UK and Germany. J Aging Stud. 2016;39:54–65.
41. Wang YC, Lee WY, Chou MY, Liang CK, Chen HF, Yeh SJ, et al. Cost and effectiveness of long-term care following integrated discharge planning: a prospective cohort study. Healthcare (Basel). 2021;9:11.
42. Schoeni RF, Freedman VA, Martin LG. Why is late-life disability declining? Milbank Q. 2008;86(1):47–89.
43. Freedman VA, Martin LG, Schoeni RF, Cornman JC. Declines in late-life disability: the role of early- and mid-life factors. Soc Sci Med. 2008;66(7):1588–602.
44. Taylor DH, Hasselblad V, Henley SJ, Thun MJ, Sloan FA. Benefits of smoking cessation for longevity. Am J Public Health. 2002;92(6):990–6.
45. Pryde MM, Kannel WB. Efficacy of dietary behavior modification for preserving cardiovascular health and longevity. Cardiol Res Pract. 2010;2011:820457.
46. Mons U, Müezzinler A, Gellert C, Schöttker B, Abnet CC, Bobak M, et al. Impact of smoking and smoking cessation on cardiovascular events and mortality among older adults: meta-analysis of individual participant data from prospective cohort studies of the CHANCES consortium. BMJ. 2015;350:h1551.
47. Beijersbergen CM, Hortobágyi T, Beurskens R, Lenzen-Grossimlinghaus R, Gäbler M, Granacher U. Effects of power training on mobility and gait biomechanics in old adults with moderate mobility disability: Protocol and Design of the Potsdam Gait Study (POGS). Gerontology. 2016;62(6):597–603.

48. O'Doherty MG, Cairns K, O'Neill V, Lamrock F, Jørgensen T, Brenner H, et al. Effect of major lifestyle risk factors, independent and jointly, on life expectancy with and without cardiovascular disease: results from the consortium on health and ageing network of cohorts in Europe and the United States (CHANCES). Eur J Epidemiol. 2016;31(5):455–68.
49. Kenfield SA, Stampfer MJ. Healthy behaviours yield major benefits in ageing. BMJ. 2013;347: f5156.
50. Olufadewa I, Adesina M, Ayorinde T. Global health in low-income and middle-income countries: a framework for action. Lancet Glob Health. 2021;9(7):e899–900.
51. Gu D, Andreev K, Dupre ME. Major trends in population growth around the world. China CDC Wkly. 2021;3(28):604–13.
52. Cardoso RSS, Tosin MHS, de Oliveira BGRB, Moraes KL, Santana RF. The multidimensionality of low health literacy in older adults: a scoping review of international studies. Clin Nurs Res. 2023;32(2):270–7.
53. Harpham T, Stephens C. Urbanization and health in developing countries. World Health Stat Q. 1991;44(2):62–9.
54. Eggleston K, Walter H. Healthy aging in Asia, vol. xxvi. Shorenstein Asia-Pacific Research Center, Freeman Spogli Institute for International Studies, Stanford University; 2020. p. 278. ISBN: 9781931368544
55. Maharaj P. Health and care in old age in Africa. London; New York: Routledge; 2020. xiii, 160 pages
56. Carmel S. Health and Well-being in late life: gender differences worldwide. Front Med (Lausanne). 2019;6:218.
57. Adeyeye SAO, Ashaolu TJ, Bolaji OT, Abegunde TA, Omoyajowo AO. Africa and the nexus of poverty, malnutrition and diseases. Crit Rev Food Sci Nutr. 2023;63(5):641–56.
58. Vaishnav LM, Joshi SH, Joshi AU, Mehendale AM. The National Programme for Health Care of the Elderly: a review of its achievements and challenges in India. Ann Geriatr Med Res. 2022;26(3):183–95.
59. Dumka N, Mangat S, Ahmed T, Hannah E, Kotwal A. Adding health to years: a review of the National Programme for Health Care of the Elderly (NPHCE) in India. J Family Med Prim Care. 2022;11(11):6654–9.
60. Hamilton-West K, Milne A, Hotham S. New horizons in supporting older people's health and wellbeing: is social prescribing a way forward? Age Ageing. 2020;49(3):319–26.
61. Wysocki A, Butler M, Kane RL, Kane RA, Shippee T, Sainfort F. Long-term services and supports for older adults: a review of home and community-based services versus institutional care. J Aging Soc Policy. 2015;27(3):255–79.
62. Asaria M, Mazumdar S, Chowdhury S, Mazumdar P, Mukhopadhyay A, Gupta I. Socioeconomic inequality in life expectancy in India. BMJ Glob Health. 2019;4(3):e001445.
63. Lama P, Claireaux H, Flower L, Harding IJ, Dolan T, Le Maitre CL, et al. Physical disruption of intervertebral disc promotes cell clustering and a degenerative phenotype. Cell Death Discov. 2019;5:154.
64. Lotz MK, Kraus VB. New developments in osteoarthritis. Posttraumatic osteoarthritis: pathogenesis and pharmacological treatment options. Arthritis Res Ther. 2010;12(3):211.
65. Foster NE, Eriksson L, Deveza L, Hall M. Osteoarthritis year in review 2022: epidemiology & therapy. Osteoarthr Cartil. 2023;31(7):876–83.
66. Li Z, Zhang Z, Ren Y, Wang Y, Fang J, Yue H, et al. Aging and age-related diseases: from mechanisms to therapeutic strategies. Biogerontology. 2021;22(2):165–87.
67. Lewis RQ, Checkland K, Durand MA, Ling T, Mays N, Roland M, et al. Integrated Care in England—what can we learn from a decade of National Pilot Programmes? Int J Integr Care. 2021;21(4):5.
68. Althaus F, Paroz S, Hugli O, Ghali WA, Daeppen JB, Peytremann-Bridevaux I, et al. Effectiveness of interventions targeting frequent users of emergency departments: a systematic review. Ann Emerg Med. 2011;58(1):41–52.e42

69. Alam M. Institute of Economic Growth (India). Ageing in India: socio-economic and health dimensions. New Delhi: Academic Foundation; 2006. p. 288.
70. Weber D, Loichinger E. Live longer, retire later? Developments of healthy life expectancies and working life expectancies between age 50-59 and age 60-69 in Europe. Eur J Ageing. 2022;19(1):75–93.
71. Nagarajan NR, Sixsmith A. Policy initiatives to address the challenges of an older population in the workforce. Ageing Int. 2023;48(1):41–77.

Chapter 5
Future Investment Needs for the Aging Community

Polly Lama , Binod Kr Tamang, and Sonam Choden Bhutia

Abstract Developing sustainable, self-sufficient, and prosperous aging communities will cut costs, create opportunities, and empower the well-being of the aging society globally. Thus, designing appropriate aged care facilities can focus on the identification of specific needs of older people. Here, we propose strategies and actions that can be adapted by countries and cities to aid the aging communities lead happy and fulfilling lives. The priorities as described have the potential to be further developed as programs and policies.

Keywords Aging · Sustainable plans · Expenditure · Healthcare · Technology

5.1 Introduction

Education, technological innovation, industrialization, medical facilities, and abundant food supplies together have increased the average life expectancy of the human population globally [1]. Financial expenditure, social policies, and healthcare program are the three main factors with a significant implication toward developing a sustainable and prosperous aging community enabling growth while addressing the challenges of the aging population [2, 3]. As described in Chaps. 1, 2, 3 and 4, the demands and needs of the elderly vary between continents and their countries socially, economically, and healthcare facility-wise the most. These differences shape the impending need to adopt better newer policies that would involve investments for restructuring the present policies, allowing development centered around societal preferences as desired by the old and the retired.

Older aging generation adapts to modernization after each passing decade increasing the average life expectancy, and in the twenty-first century the elders pursue a much active, technologically connected life than the previous generations,

P. Lama (✉) · B. K. Tamang · S. C. Bhutia
Sikkim Manipal Institute of Medical Sciences, Sikkim Manipal University, Gangtok, Sikkim, India
e-mail: pollylama@smims.smu.edu.in

P. Lama (ed.), *The Ageing Population*,
https://doi.org/10.1007/978-981-99-5772-9_5

and the evidences from Chaps. 2, 3 and 4 suggest that to an extent the elder are experiencing better lives that fulfills certain aspects of healthy aging process. Thus, it can be expected that older people will continue to live longer in the future due to rapid advancement in healthcare, technology, and better social-economic welfare plans. Older adults today, either out of need or by choice, can remain employed in a full or part-time capacity, and some even pursue new careers, and many are engaged with volunteer tasks. Rapid technological development especially during the COVID-19 phase is perhaps one of the main factors behind such progressive change, reshaping jobs, education, training, and social outreach. However, in low-income countries longevity also means an increase in the number of uneducated less qualified older adults who suffer the most due to lack of implementation of literacy tools, and as described in the previous chapters, educational disadvantage directly correlates with poorer health choices and greater incidence of non-communicable and communicable diseases affecting the quality of life and other opportunities, described in detail in Chaps. 1, 2 and 3, thus better physical and mental health means investment, growth and adequate financial contentment.

Therefore, the main focus behind this chapter is to present attainable practical priorities that can adapt to demographic changes while addressing the critical demands of the aged, retired, and the frail oldest old. These priorities have the potential to be further developed as programs and policies, and are discussed under the following themes:

1. Strengthening the health infrastructure.
2. Provision for ending housing needs and nutritional deficits.
3. Investing in education and lifelong learning.
4. Utilizing renewable and sustainable sources of energy in eldercare.

5.1.1 Strengthening the Healthcare Infrastructure

As the aging population grows, it is almost undeniable that the use of health services will also rise, which adds to the current cost and is set to increase gradually over the next decade [4]. In the less advanced countries, the demand for providing better healthcare services lack severely, especially within the rural regions of many countries, still in the process of setting up primary care centers/hospitals with adequate facilities [5]. Focused geriatric care services are non-existent in less advanced low, and middle-income countries, and the dependent oldest old are supported by their own families or a few aided organizations operating as the old-age home, which can be costly, as long-term care (LTC) for the oldest old, and the disabled increases manpower needs and expenditure for the healthcare system. Thus, aged care facilities can be improved around three services: homecare-based support, sheltered housing or old age homes, and nursing homes for the frail.

"*Home care-based support*" can be administered by the family members who may be trained by an expert for elder care, or through members involved with community care support programs, however, the cost associated with providing these training services may require careful considerations [6]. These types of services are often available through payment by private bodies in a few low, and middle-income countries. In high-income countries, for example, Australia, the United Kingdom, and Denmark, home care services by an expert are often free to access facility for the oldest old, while some of the European countries have adequate financial support to pay for expert care, however, shortages are experienced in terms of trained manpower in geriatric care [7, 8].

"*Sheltered housing or old age homes*" refers to a grouped housing "scheme" of flats or homes exclusively for older people with a manager or in charge living with the older residents. These types of services are present in high, as well as low-income countries, but are predominantly limited to the urban regions in low- and middle-income countries [9]. Sheltered accommodation in rich countries varies and can be self-contained, ranging from a simple bedsit to a large flat or a house, with expert caregivers, and a doctor on call facilities [10]. This scheme is different from a nursing home, as the older residents residing in old age homes are usually active with a degree of independence, being able to communicate about their medical needs to expert caregivers [11]. These housing facilities also have community centers that conduct educational and recreational programs on a regular basis. Sheltered housing schemes are generally owned, run, and maintained by local authorities or housing associations in high-income countries, and in low, and middle-income countries such facilities are available in urban regions, and are maintained by private bodies, NGOs, and a few are Government facilities.

"*Nursing homes*" are services offered to the oldest old who suffer from one or more chronic health complications and require constant medical monitoring or delivery of therapeutics through trained nurses. Nursing homes for the mentally ill suffering from common old-age conditions such as dementia, Alzheimer's, and Parkinsonism are kept in protected facilities with trained experts to care for them thus the cost for care, whether provided by the Government or private body, will be higher for nursing home care, and existing manpower may not be sufficient as the proportion of terminally frail and dependent group of older adults may increase as the aging population increases.

5.1.2 Provision for Ending "Housing" Need and "Nutritional" Deficit

As the aging population increases demand for a new segment of "housing need" and "adequate nutrition" may arise to fulfill the prospect of healthy aging. Newer housing needs can be addressed with features such as accessibility, technology-assisted care services while taking in account the health conditions, family structure, culture, and income level of the elderlies. Numerous housing concepts can be designed ranging from independent living with periodic expert care assistance, co-housing, shared housing, and multigenerational housing with care support on a weekly basis, to retirement communities/villages/cities with residing retired medical practitioners or nurses [12].

In developed and advanced countries, residential independence is sought by the elderly aged between 65 and 80 years, while in developing and less advanced countries, traditionally the elderly live with their adult children, but this practice is declining due to longer life expectancy, better health conditions, financial independence, and preference to live on their own until at a much later stage [13]. In developed countries, the elderly who live on their own also have better access to communal services, transport, and health care facilities. This type of system is yet to be conceived in less advanced countries.

Housing needs in the future may have to focus more on the most vulnerable oldest old group over 80 years, as increase in their numbers would mean they will require new and adjusted housing solutions for their specific needs. Therefore, widening the range of dwellings designed to suit the needs of the oldest old will perhaps become a practical approach toward long-term care. An economically fragile and socially disadvantaged group of oldest old may burden the housing needs and is likely to be subjected to discrimination if proper measures are not adopted.

"*Nutrition*" declines with age due to decrease consumption of food as the older adults frequently report "loss in appetite." Changes in physical and physiological factors may decrease energy levels and perhaps can be one of the possible explanation for the decreased food intake as age progresses [14]. However, it is also possible that biological changes in hunger pathways with age may lead to decrease consumption, but similar studies are rarely reported with thirst which is essential to define an accurate reason. It is also suggested that consumption of a diet depends partly on sensory-specific brain satiety pathways, that decrease the pleasantness of a food consumed with growing age [15], and this could perhaps be the reason why some elderly consume more monotonous diets. Thus, decrease food intake will influence the nutritional requirement of the body. Lack of nutritious diet may lead to vitamin deficiency diseases that can alter the metabolic processes of the body. Thus, it is necessary to provide elderlies above 80 years with a well-planned healthy diet (Fig. 5.1).

5.1.3 Investing in Education and Lifelong Learning

Education at older ages is an important tool that can bring demographic changes toward economic growth and productivity. The existing aging labor forces will require considerable improvement efforts updating newer trends with technology-based education so as to increase the employability of the older workers, and this is essential to counteract early or after retirement job prospects and for involving older adults with less labor-intensive computer/technology-assisted desk tasks. Similarly,

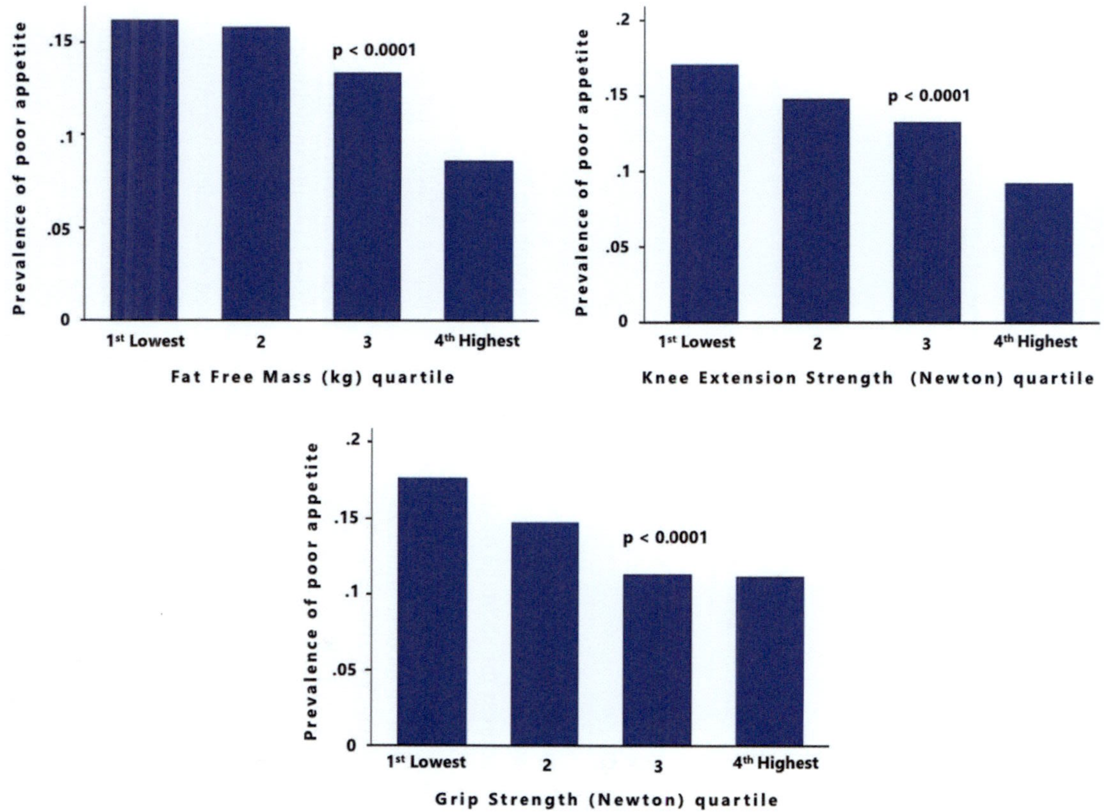

Fig. 5.1 Prevalence of poor appetite or ability to eat with age (1000+ subjects) with changes to body composition and muscle strength. Cut-off points for quartile levels of each variable: *Fat-free mass* (kg, <44.5, 44.6–52.1, 52.2–62.2, 62.3+); *Knee extension strength* (*N*, <231.8, 231.9–300.3, 300.4–390.4, 390.5+); *Grip Strength* (*N*, <215.4, 215.8–274.6, 274.7–367.4, 367.5+). The prevalence of poor appetite or ability to eat was highest in the lowest quartiles of fat free mass, knee extension strength and grip strength [16]

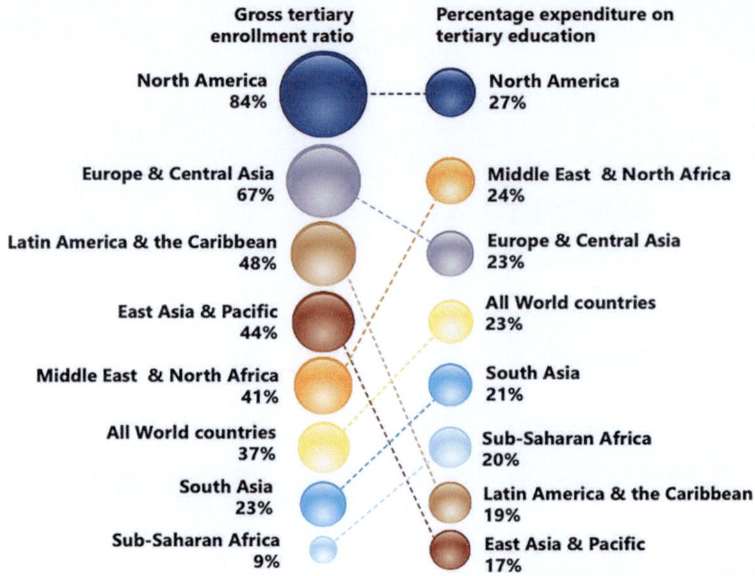

Fig. 5.2 Regional variation in gross tertiary education enrollment ratio and percentage expenditure, World Bank Development Report, 2016 [17]

training the aging professionals dedicated to elderly care with the latest skill-based practices to address the shortage of staff will additionally be a vital step toward care of the oldest old, which is one of the fastest growing employment prospects globally at the current phase of societal development. These mentioned factors for healthy aging will incur a modest financial expenditure but will offer productivity and promote employability (Fig. 5.2).

Development of a lifelong learning approach may involve implementing policies to employ older adults after retirement and competitiveness for the individuals to seek training at a later stage, and for the firms/organization to offer it. Thus, increasing investment in public and private resources to deliver a lifelong learning attitude remains a challenge, particularly for less educated older adults, especially in the event of economic and the recent pandemic-afflicted crisis.

Updating skills during working life to respond to the changing needs of the labor market is critical if older workers are to remain at work longer, and lifelong learning is a key factor for employment opportunities. Lifelong learning and training if adopted rigorously can also play a role in overcoming the lack of formal education or acquisition of basic skills for the less educated elderlies and help them become better adapted to newer needs of the labor market.

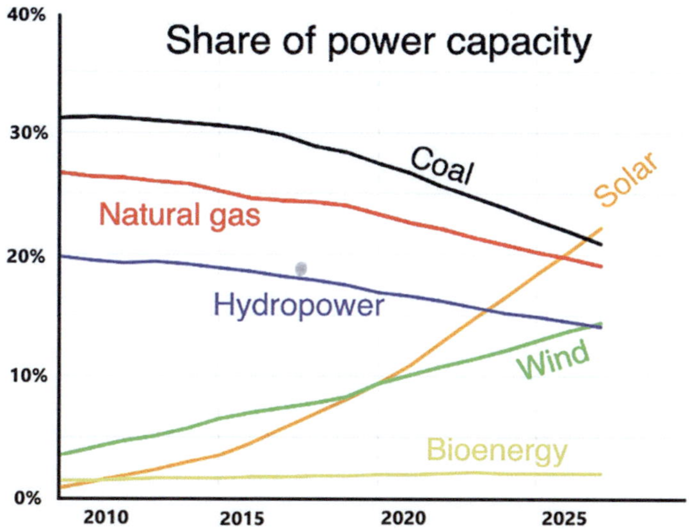

Fig. 5.3 Fossil fuel capacities in the development of sustainable energy sources [3]

5.1.4 *Utilizing Renewable and Sustainable Sources of Energy in Eldercare*

Using ecofriendly, renewable, and biodegradable sources of energy to sustain the existing needs for power consumption is a factor that is applicable to everyone, but there are large differences across continents and countries with the prospect of utilizing and harnessing these energy sources as economic status, organizational and residential sector priorities, landscape, and climatic conditions vary between countries and continents. Thus, investments to be made should recognize diversities concerning socioeconomic viability, institutional, and geographical conditions before addressing the cost of implementation of newer strategies.

An agenda of the United Nations' sustainable development goals is to interlink human development, use of ecofriendly measures, and social objectives together to foster cross-cutting integration of themes promoting economic ventures. Therefore, projects associated with the use of renewable and sustainable reforms may involve developing energy saving ideas and approaches, exploring thermal rehabilitation, heating system-built form available natural gases, rainwater harvesting system, and solar powered electrical/lighting system, photovoltaic energy system, biomass, geothermic or wind devices for electricity generation purposes. Green building construction can be encouraged to cooperate thermal insulation of walls, roofs, and windows for efficient heating and cooling needs, use of passive lighting and active shading can be introduced while designing new houses, use of energy-efficient appliances, for example, LED lighting to reduce power consumption, use of recycled, reused, or low-energy building materials are all efficiency measures appliable globally to all ages (Fig. 5.3).

Competing Interest The authors have no conflicts of interest to declare that are relevant to the content of this chapter.

References

1. Wirayuda AAB, Chan MF. A systematic review of sociodemographic, macroeconomic, and health resources factors on life expectancy. Asia Pac J Public Health. 2021;33(4):335–56.
2. Organisation for Economic Co-operation and D. Working better with age. Paris, France: OECD; 2019.
3. United N. World population prospects 2022 : summary of results. New York: United Nations; 2022.
4. Gu D, Andreev K, Dupre ME. Major trends in population growth around the world. China CDC Wkly. 2021;3(28):604–13.
5. United Nations. Economic and Social Commission for Asia and the Pacific. Population division. Ageing and the family in the developing Asian and Pacific countries. New York, NY: United Nations; 1994. p. 73–5.
6. Roquebert Q, Tenand M. Informal care at old age at home and in nursing homes: determinants and economic value. Eur J Health Econ. 2023;
7. Wysocki A, Butler M, Kane RL, Kane RA, Shippee T, Sainfort F. Long-term services and supports for older adults: a review of home and community-based services versus institutional care. J Aging Soc Policy. 2015;27(3):255–79.
8. Calderón-Jaramillo M, Zueras P. Cared and uncared populations: understanding unmet care needs of older adults (65+) across different social care systems in Europe. Eur J Ageing. 2023;20(1):11.
9. Piya N, Shah T, Badhu A, Shrestha S. Comparision of quality of life of senior citizens residing in old age homes and own residence. J Nepal Health Res Counc. 2020;18(1):16–20.
10. Song P, Chen Y. Public policy response, aging in place, and big data platforms: creating an effective collaborative system to cope with aging of the population. Biosci Trends. 2015;9(1): 1–6.
11. Tinker AM. Housing for frail elderly people. Public Health. 1992;106(4):301–5.
12. Green G. Age-friendly cities of Europe. J Urban Health. 2013;90(Suppl 1):116–28.
13. Van Hoof J, Kazak JK, Perek-Białas JM, Peek STM. The challenges of urban ageing: making cities age-friendly in Europe. Int J Environ Res Public Health. 2018;15:11.
14. Drewnowski A, Shultz JM. Impact of aging on eating behaviors, food choices, nutrition, and health status. J Nutr Health Aging. 2001;5(2):75–9.
15. Spencer SJ, Korosi A, Layé S, Shukitt-Hale B, Barrientos RM. Food for thought: how nutrition impacts cognition and emotion. NPJ Sci Food. 2017;1:7.
16. Chang M, Geirsdottir OG, Launer LJ, Gudnasson V, Visser M, Gunnarsdottir I. A poor appetite or ability to eat and its association with physical function amongst community-dwelling older adults: age, gene/environment susceptibility-Reykjavik study. Eur J Ageing. 2021;18(3): 405–15.
17. World Bank (Washington District of Columbia). World Development Report 2021: data for better lives. Washington: World Bank. p. 2021.

Chapter 6
Variabilities in Ageing Trends: Questions and Concluding Remarks

Polly Lama

Abstract The concept behind ageing gracefully and increasing longevity are perhaps the two essential components behind redefining well-being and setting up newer policies for the 'ageing society'. Approximately 80% of the countries in the World have surpassed the threshold for being referred to as 'rapidly ageing'. These changes in ageing trend are due to put pressure on many current plans and policies and will question the strength of healthcare expenditure and pension plans the most. Lesson learned from COVID-19 pandemic, increasing proportion of elderlies, and age-related constraints therefore warrants some critical questions which is discussed in this conclusive overview.

Keywords Ageing · Ageing society · Healthcare · Manual labour · Income

6.1 Introduction

A common numerical benchmark used to denote an aging society is when 17% of the population in high-income countries, 13% in middle income, and 7% in low-income countries reach the age of 65 years and above. As described in Chap. 1, almost 80% of the countries have already surpassed the cut-off for being labelled as an ageing society, and undeniably the World population is ageing rapidly. A few countries reached the cut-off value faster than ever before in the recent past mainly due to increase in life expectancy, thereby increasing the old-age dependency ratio. These changes will impose pressure on the smaller working-age group productive population, and utilisation of public budgets for old-age reforms may increase given the rising need to pay for long-term care services for the frail oldest old, unable to work. However, these changes are gradual but progressive, and new policies can be planned for elderly care by carefully appraising the current and proposed welfare

P. Lama (✉)
Sikkim Manipal Institute of Medical Sciences, Sikkim Manipal University, Gangtok, Sikkim, India
e-mail: pollylama@smims.smu.edu.in

P. Lama (ed.), *The Ageing Population*,
https://doi.org/10.1007/978-981-99-5772-9_6

regime, the institutional capacities, and regulatory framework of each country in operation to meet the expenditure.

For instance, in low- and middle-income countries, where large portions of the population tend to work in the agricultural or manual labour sector, old-age social security system has limited coverage, and older people may remain longer in labour force because when they stop working, they may be fully reliant on their families for financial support. Thus, instilling educational reforms at early ages and promoting literacy amongst older ages may be necessary to enhance re-employment opportunities. Old-age homes may also become a priority if the family care culture begins to diminish, thus adopting suitable insurance/financial schemes specially designed for manual/agricultural labour may have a positive outcome for the aged.

An effective and active aging strategy can in-cooperate the contributions of the citizens and the country/region objectively, to promote cross-cutting integrations of independent sectors such as employment, healthcare, social protection, pensions, social inclusion, technology, and economic policy [1]. The translation of this approach into concrete policies would amend the current system as one entity binding employment prospect, expenditure, and health reforms, to prevent the loss of contributions and skills acquired in the latter part of the life course due to lack of physical working capacities. The redistribution of existing resources at an early preventative stage is, therefore, an essential criterion, whether it is the pension policies or the healthcare system [2], as newer practices should not create a barrier between employment, earnings, retirement, and should facilitate post-retirement participation [3]. Underpinning this belief is the fact that chronological age is not an effective predictor of performance and experience gained during one's life, no matter the vocation, will have a contributory effect. A life course planning perspective thus will reinforce the preventative dimension of ageing gracefully, and in this regard ageing well will require strategies to promote health, well-being, and financial planning prior to transformation as the oldest old [4]. The new healthy ageing policies should facilitate to seek the maintenance of equilibrium between physical, mental, and financial capacities, which would enable older adults to work longer if necessary, and will create healthier and more fulfilled post-employment years [5].

Thus, as the proportions of the older grows, many new situations will emerge in terms of caregiving and fulfilling the basic needs for some of the economically or health-wise compromised aged group.

The facts presented in this book also pose questions particularly if healthy ageing is not achieved. Especially in the case of the Asian and African countries where it is likely that older aged adults may have to remain in the labour force longer compared to adults from other world regions, so would this mean forced labour will continue to rise as the aged group accumulates?

The status of pension system is extensively discussed with relevance to countries and continents, thus what remains unanswered is to what extent would the pension scheme be able to reduce the reliance on earnings and other resources provided by the families?

The challenges associated with the current healthcare system particularly centres around the rise in communicable and non-communicable diseases after a certain age as the frail oldest old are more susceptible to immunosenescence favouring a sharp decline in immune system function after the age of 75 years, and difference in health care infrastructure between various countries pose another questions for the older adults in less advance countries as to are they really living longer and healthier than other older adults in more advanced World regions?

The common causes behind mortality and morbidity in the World are presented in Chap. 2, and what still remains unanswered is would the current healthcare system be enough prepared to tackle health-related issues if another epidemic or pandemic hits the World at large?

As extensively discussed in Chaps. 1–5, the availability, accessibility, and affordability of adequate infrastructure and services for the elderly will be crucial to ensuring decent living conditions. But, what still remains unclear is whether the current provision for healthcare infrastructure, age releated reform policies and services be adequate to forestall global challenges for older adult care and can the advanced and less advance countries meet the longer-term and demand-driven perspectives as the ageing population keeps growing?

Looking ahead, it is at least clear that each individual country and continent will need to unite and adjust to realities, for example the situation as posed by COVID-19 pandemic that particularly highlighted the shortcomings in medical care infrastructure around the World claiming more lives of the older adults. The brutal fact that discrepancies exist between older citizens obtaining appropriate care and services depend on the economic profile of the region questions the costs associated with healthy ageing and the desire for independence over institutionalisation. Most importantly, healthcare system, need to address ageism with a more practical and evidence-based cost-effective approach.

Ageing gracefully is thus an opportunity to explore newer ways to build the economy and the World with a productive approach. One of the most promising aspects related to ageing is perhaps the application of cellular and molecularly targeted anti-ageing treatments to cure disorders associated with ageing. However, the current research evidence indicates that such therapies are still at an incipient stage.

Thus, this book has summarised and represented the details about the current societal standing and healthcare cost influencing the quality of life for the 'Ageing World'.

Competing Interest The author has no conflicts of interest to declare that are relevant to the content of this chapter.

References

1. Walker A, Foster L, Edward EP. The political economy of ageing and later life: critical perspectives. Cheltenham: Edward Elgar Pub. Ltd; 2014. 1 online resource (v.) p
2. Organisation for Economic Co-operation and D. Working better with age. Paris, France: OECD; 2019.

3. Greer SL. Ageing and health : the politics of better policies. New York, NY: Cambridge University Press; 2021. pages cm. p
4. Skirbekk V, Dieleman JL, Stonawski M, Fejkiel K, Tyrovolas S, Chang AY. The health-adjusted dependency ratio as a new global measure of the burden of ageing: a population-based study. Lancet Healthy Longev. 2022;3(5):e332–e8.
5. Bloom DE, Chatterji S, Kowal P, Lloyd-Sherlock P, McKee M, Rechel B, et al. Macroeconomic implications of population ageing and selected policy responses. Lancet. 2015;385(9968): 649–57.

Appendix: World Population Data Sheet

P. Lama (ed.), *The Ageing Population*,
https://doi.org/10.1007/978-981-99-5772-9

SPECIAL FOCUS ON THE DEMOGRAPHIC IMPACTS OF COVID-19

COVID-19 Pandemic Has Fueled Excess Deaths

Excess deaths refers to deaths occurring directly or indirectly because of the pandemic, including deaths from COVID-19 itself and from delayed or unavailable health care. It also accounts for decreases in deaths from other causes, such as traffic accidents and influenza, due to changes in people's behaviors. Excess deaths vary across regions and countries, reflecting variations in population age structures; health infrastructures; COVID-19 containment measures, including vaccination rates; and social and economic conditions.

Excess Deaths Due to the Pandemic as a Percent of Total Deaths, 2020-2021

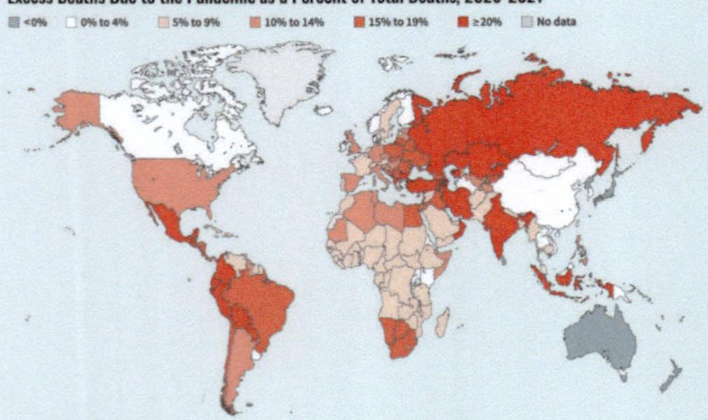

Note: Data reflect 2020-2021 annual average.
Source: World Health Organization, Global Excess Deaths Associated With COVID-19 (modelled estimates) data set.

Pandemic Effects on Fertility Are Largely Limited

High-income countries like Germany and the United States mostly saw small declines in births in 2020 that rebounded or stabilized in 2021. Low- and middle-income countries such as Costa Rica and Türkiye continued to see births decline, following pre-pandemic trends. These data suggest the pandemic's impact on fertility has generally been limited and temporary.

Births per 1,000 Population, 2018-2021

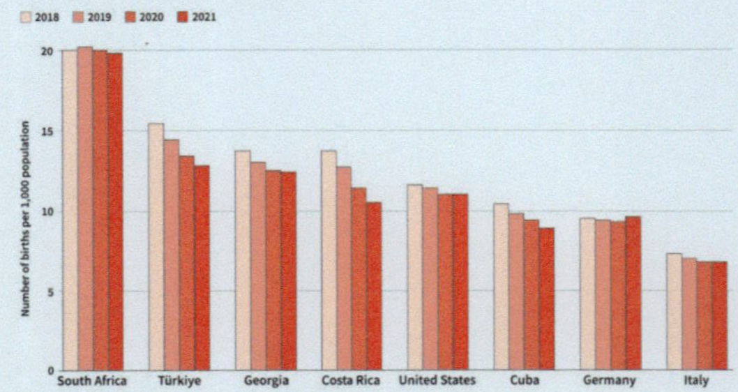

Source: National Statistical Offices from listed countries.

WORLD

	Population (millions) mid-2022	Births per 1,000 Population	Deaths per 1,000 Population	Rate of Natural Increase (%)	Net Migration Rate	Population (millions)		Infant Mortality Rate[a]	Total Fertility Rate[b]	Percent of Population		Family Planning Among Married Women Ages 15-49[c]			Life Expectancy at Birth (Years)			Population Ages 15-49 With HIV/AIDS (%)		Urban Population (%)	Population per km² of Arable Land	GNI per Capita, PPP[e]	SPECIAL FOCUS COVID-19 INDICATORS: Excess Deaths Due to COVID-19 Pandemic, 2020-2021 Annual Average[f]			Population Fully Vaccinated Against COVID-19 (%)[g]
						mid-2035	mid-2050			Ages <15	Ages 65+	Using All Methods (%)	Using Modern Methods (%)	Demand Satisfied by Modern Methods (%)[d]	Total	Males	Females	Males	Females				No. of Excess Deaths	Excess Deaths per 100,000 Population	Excess Deaths as % of Total Deaths	
WORLD	7,963	17	8	0.9	—	8,899	9,752	29	2.3	25	10	63	55	—	72	70	75	0.6	0.8	57	2,168	18,625	7,455,097	96	12	62
More Developed	1,270	10	12	-0.2	2	1,298	1,297	4	1.5	16	20	67	57	—	78	75	82	—	—	79	685	50,865	1,824,632	142	11	70
Less Developed	6,694	19	7	1.2	-0	7,601	8,454	31	2.4	27	8	62	55	—	71	69	73	—	1.1	53	2,450	12,199	5,721,468	87	11	60
Least Developed	1,126	31	8	2.4	-0	1,489	1,935	43	4.0	39	4	39	35	54	64	62	67	1.0	1.7	35	1,004	3,399	503,787	47	7	—
High Income	1,241	10	10	0.0	2	1,289	1,310	4	1.5	16	19	66	—	—	80	77	83	—	—	81	8,430	55,066	1,127,703	90	9	74
Middle Income	5,953	17	8	0.9	-0	6,553	7,030	27	2.1	25	9	65	58	—	72	69	74	—	—	54	1,036	12,909	5,985,461	102	13	64
Upper-Middle Income	2,527	10	8	0.2	-0	2,588	2,555	10	1.5	20	13	77	71	—	76	73	79	—	—	70	984	19,622	2,053,784	82	10	77
Lower-Middle Income	3,427	22	7	1.5	-0	3,964	4,475	34	2.6	29	6	57	49	69	68	67	70	0.5	0.7	43	1,074	7,904	3,931,678	118	15	54
Low Income	738	35	8	2.7	-0	1,020	1,372	46	4.6	42	3	34	30	47	62	60	65	1.1	1.9	34	754	2,135	330,795	47	7	17

NORTHERN AFRICA

2022 WORLD POPULATION DATA SHEET | 3

	Population (millions) mid-2022	Births per 1,000 Population	Deaths per 1,000 Population	Rate of Natural Increase (%)	Net Migration Rate	Population (millions)		Infant Mortality Rate[a]	Total Fertility Rate[b]	Percent of Population		Family Planning Among Married Women Ages 15-49[c]			Life Expectancy at Birth (Years)			Population Ages 15-49 With HIV/AIDS (%)		Urban Population (%)	Population per km² of Arable Land	GNI per Capita, PPP[e]	SPECIAL FOCUS COVID-19 INDICATORS: Excess Deaths Due to COVID-19 Pandemic, 2020-2021 Annual Average[f]			Population Fully Vaccinated Against COVID-19 (%)[g]
						mid-2035	mid-2050			Ages <15	Ages 65+	Using All Methods (%)	Using Modern Methods (%)	Demand Satisfied by Modern Methods (%)[d]	Total	Males	Females	Males	Females				No. of Excess Deaths	Excess Deaths per 100,000 Population	Excess Deaths as % of Total Deaths	
AFRICA	1,419	33	9	2.4	-0	1,891	2,478	47	4.3	40	3	36	32	51	63	61	64	1.8	3.4	44	894	5,212	820,657	61	8	20
NORTHERN AFRICA	**251**	**23**	**6**	**1.6**	**-0**	**301**	**356**	**22**	**2.8**	**33**	**5**	**50**	**46**	**65**	**73**	**71**	**75**	**0.1**	**0.1**	**53**	**1,590**	**10,594**	**212,559**	**87**	**13**	**31**
Algeria	44.9	22	5	1.7	-0	52.4	60.0	19	2.9	31	6	54	45	66	76	75	78	<0.1	<0.1	74	598	11,750	34,794	79	14	16
Egypt	103.5	21	7	1.4	1	122.7	144.8	16	2.5	34	4	*59*	*57*	*80*	75	73	76	<0.1	<0.1	44	3,077	12,910	125,551	122	18	35
Libya	6.8	17	6	1.1	-0	7.7	8.5	9	2.4	28	5	*28*	*16*	*24*	72	70	75	0.2	0.1	81	396	23,550	3,931	57	12	17
Morocco	36.7	17	5	1.2	-1	40.7	43.6	15	2.1	25	8	71	59	72	74	72	76	<0.1	<0.1	64	479	8,020	17,374	47	6	63
Sudan	46.9	34	7	2.7	-0	63.6	84.5	39	4.5	41	4	*12*	*12*	*30*	65	63	68	0.2	0.1	36	223	4,000	19,000	43	7	9
Tunisia	11.8	16	8	0.8	-1	12.9	13.7	14	2.1	25	9	51	44	63	74	71	77	<0.1	<0.1	70	456	11,270	11,911	100	14	54
Western Sahara	0.6	17	6	1.1	6	0.7	0.9	26	2.3	24	6	—	—	—	71	69	73	—	—	87	14,400	—	—	—	—	—

WESTERN AFRICA

	Population (millions) mid-2022	Births per 1,000 Population	Deaths per 1,000 Population	Rate of Natural Increase (%)	Net Migration Rate	Population (millions) mid-2035	Population (millions) mid-2050	Infant Mortality Rate[a]	Total Fertility Rate[b]	Percent of Population Ages <15	Percent of Population Ages 65+	Family Planning Among Married Women Ages 15-49[c]: Using All Methods (%)	Using Modern Methods (%)	Demand Satisfied by Modern Methods (%)[d]	Life Expectancy at Birth (Years): Total	Males	Females	Population Ages 15-49 With HIV/AIDS (%): Males	Females	Urban Population (%)	Population per km² of Arable Land	GNI per Capita, PPP[e]	SPECIAL FOCUS COVID-19 INDICATORS — Excess Deaths Due to COVID-19 Pandemic, 2020-2021 Annual Average[f]: No. of Excess Deaths	Excess Deaths per 100,000 Population	Excess Deaths as % of Total Deaths	Population Fully Vaccinated Against COVID-19 (%)[g]
AFRICA	**1,419**	**33**	**9**	**2.4**	**-0**	**1,891**	**2,478**	**47**	**4.3**	**40**	**3**	**36**	**32**	**51**	**63**	**61**	**64**	**1.8**	**3.4**	**44**	**894**	**5,212**	**820,657**	**61**	**8**	**20**
SUB-SAHARAN AFRICA	**1,168**	**35**	**9**	**2.6**	**-0**	**1,591**	**2,122**	**50**	**4.6**	**42**	**3**	**33**	**29**	**47**	**60**	**58**	**62**	**2.2**	**4.2**	**42**	**739**	**3,993**	**608,098**	**55**	**7**	**17**
WESTERN AFRICA	**430**	**36**	**10**	**2.5**	**-0**	**583**	**773**	**60**	**4.9**	**43**	**3**	**20**	**16**	**37**	**57**	**57**	**58**	**0.8**	**1.7**	**48**	**627**	**4,397**	**188,016**	**46**	**6**	**12**
Benin	13.4	36	9	2.7	-0	18.5	25.3	54	4.9	42	3	16	12	26	60	58	62	0.5	1.1	49	477	3,750	5,927	48	6	22
Burkina Faso	22.7	35	9	2.6	-1	30.8	40.5	47	4.7	44	3	35	32	56	60	58	62	0.5	0.8	31	378	2,330	11,080	52	7	7
Cabo Verde	0.6	15	5	0.9	-2	0.7	0.7	11	1.8	26	6	56	55	—	74	70	78	0.6	0.6	74	1,186	6,890	464	83	13	—
Côte d'Ivoire	28.2	34	9	2.5	-1	38.3	51.4	57	4.4	41	2	28	22	41	59	57	60	1.2	2.7	52	805	5,760	12,717	48	6	21
Gambia	2.7	33	8	2.6	-1	3.6	4.7	31	4.7	43	2	19	17	40	62	61	64	1.3	2.1	63	615	2,370	1,452	59	9	15
Ghana	33.5	28	8	2.0	-0	42.2	52.2	32	3.6	37	4	31	25	—	64	62	66	1.0	2.4	58	1,332	6,020	10,454	33	5	21
Guinea	13.9	34	10	2.4	-0	18.4	23.7	64	4.4	42	3	11	11	32	59	58	60	0.9	2.0	37	447	2,540	6,265	47	5	26
Guinea-Bissau	2.1	31	9	2.2	-1	2.7	3.4	51	4.0	40	3	21	20	48	60	57	62	2.2	4.0	45	702	2,060	1,421	71	8	17
Liberia	5.3	31	9	2.2	-2	6.9	8.9	58	4.1	41	3	25	24	41	61	59	62	0.8	1.4	53	1,061	1,460	2,022	39	5	28
Mali	22.6	42	9	3.2	-1	33.1	47.4	55	6.0	47	2	17	16	40	59	58	60	0.7	0.9	45	352	2,370	13,656	66	8	6
Mauritania	4.7	39	7	3.2	1	6.8	9.6	41	5.2	42	3	14	13	28	65	63	66	0.3	0.3	56	1,185	5,530	3,474	74	12	29
Niger	26.2	45	8	3.7	0	42.0	67.0	40	6.7	49	2	14	13	35	62	61	63	0.1	0.2	17	148	1,330	17,170	70	8	6
Nigeria	218.5	37	12	2.4	-0	291.6	377.5	72	5.1	43	3	17	12	34	54	53	54	0.9	1.8	53	624	5,250	93,217	45	5	8
Senegal	17.9	32	5	2.7	-1	23.9	30.6	33	4.4	41	3	27	26	53	70	68	71	0.3	0.4	49	560	3,690	8,252	49	9	9
Sierra Leone	8.7	33	10	2.3	2	11.4	14.4	75	3.9	39	3	21	21	45	58	57	60	1.0	1.8	43	549	1,750	3,943	49	6	21
Togo	8.8	32	8	2.4	-0	11.7	15.5	44	4.3	40	3	24	21	37	62	61	62	1.2	2.5	43	334	2,390	-3,495	-42	-7	19

EASTERN AFRICA

	Population (millions) mid-2022	Births per 1,000 Population	Deaths per 1,000 Population	Rate of Natural Increase (%)	Net Migration Rate	Population (millions) mid-2035	Population (millions) mid-2050	Infant Mortality Rate[a]	Total Fertility Rate[b]	Percent of Population Ages <15	Percent of Population Ages 65+
AFRICA	**1,419**	**33**	**9**	**2.4**	**-0**	**1,891**	**2,478**	**47**	**4.3**	**40**	**3**
SUB-SAHARAN AFRICA	**1,168**	**35**	**9**	**2.6**	**-0**	**1,591**	**2,122**	**50**	**4.6**	**42**	**3**
EASTERN AFRICA	**473**	**33**	**7**	**2.6**	**-0**	**643**	**851**	**41**	**4.3**	**41**	**3**
Burundi	12.9	35	7	2.7	-1	17.8	24.2	37	5.1	46	2
Comoros	0.8	29	9	2.1	-2	1.0	1.2	46	4.0	38	4
Djibouti	1.1	22	9	1.3	1	1.3	1.5	39	2.8	30	5
Eritrea	3.7	29	7	2.2	-4	4.7	6.0	29	3.9	39	4
Ethiopia	123.4	32	7	2.6	-0	165.7	214.8	47	4.2	40	3
Kenya	54.0	28	8	2.0	-1	69.0	85.2	30	3.3	38	3
Madagascar	29.6	35	7	2.8	-0	40.9	56.2	47	4.3	39	3
Malawi	20.4	33	7	2.6	-0	28.0	37.2	28	3.9	43	3
Mauritius	1.3	10	11	0.0	-1	1.3	1.2	13.8	1.4	17	12
Mayotte	0.3	34	3	3.1	7	0.4	0.6	*4*	4.4	44	3
Mozambique	33.0	37	9	2.8	-0	46.3	63.0	52	4.6	44	3
Réunion	0.9	16	7	0.9	-6	1.0	1.0	7	2.5	22	14
Rwanda	13.8	30	6	2.4	-0	18.0	23.0	29	3.8	38	3
Seychelles	0.1	17	9	0.8	0	0.1	0.1	*10.8*	2.5	22	12
Somalia	17.6	44	12	3.2	-1	25.6	36.5	71	6.3	47	3
South Sudan	10.9	29	11	1.8	-3	13.9	17.5	64	4.5	44	3
Tanzania	65.5	36	6	3.0	-0	93.1	129.9	34	4.7	43	3
Uganda	47.2	37	6	3.1	1	65.6	87.6	30	4.6	45	2
Zambia	20.0	34	7	2.7	-0	27.8	37.5	38	4.2	43	2
Zimbabwe	16.3	31	9	2.1	-2	21.0	26.4	37	3.5	41	3

	Family Planning Among Married Women Ages 15-49[c]: Using All Methods (%)	Using Modern Methods (%)	Demand Satisfied by Modern Methods (%)[d]	Life Expectancy at Birth (Years): Total	Males	Females	Population Ages 15-49 With HIV/AIDS (%): Males	Females	Urban Population (%)	Population per km² of Arable Land	GNI per Capita, PPP[e]	SPECIAL FOCUS COVID-19 INDICATORS — Excess Deaths Due to COVID-19 Pandemic, 2020-2021 Annual Average[f]: No. of Excess Deaths	Excess Deaths per 100,000 Population	Excess Deaths as % of Total Deaths	Population Fully Vaccinated Against COVID-19 (%)[g]
AFRICA	**36**	**32**	**51**	**63**	**61**	**64**	**1.8**	**3.4**	**44**	**894**	**5,212**	**820,657**	**61**	**8**	**20**
SUB-SAHARAN AFRICA	**33**	**29**	**47**	**60**	**58**	**62**	**2.2**	**4.2**	**42**	**739**	**3,993**	**608,098**	**55**	**7**	**17**
EASTERN AFRICA	**46**	**41**	**—**	**63**	**61**	**66**	**2.3**	**4.1**	**29**	**904**	**2,692**	**184,612**	**41**	**6**	**20**
Burundi	29	23	39	62	60	64	0.7	1.2	14	1,074	800	4,669	39	6	0
Comoros	*19*	*14*	*27*	63	61	66	<0.1	<0.1	30	1,268	3,260	362	41	7	35
Djibouti	*19*	*18*	—	62	60	65	0.6	0.9	78	56,042	5,740	884	89	11	15
Eritrea	*8*	*7*	*20*	67	64	69	0.4	0.6	42	534	—	1,207	34	4	—
Ethiopia	41	41	—	65	62	68	0.6	1.1	22	762	2,590	51,748	44	8	19
Kenya	65	61	77	61	59	64	2.6	5.4	28	932	4,950	6,176	11	2	17
Madagascar	50	43	66	65	63	68	0.3	0.4	39	988	1,600	12,791	46	7	4
Malawi	66	65	80	63	59	66	5.7	9.4	18	567	1,630	8,557	44	7	8
Mauritius	*64*	*32*	*42*	74	71	78	1.5	1.2	41	1,684	25,530	471	37	4	74
Mayotte	—	—	—	73	73	74	—	—	46	1,768	—	—	—	—	—
Mozambique	*27*	*25*	*50*	59	56	62	—	—	38	584	1,310	21,366	67	7	68
Réunion	—	—	—	80	77	83	—	—	100	2,569	—	—	—	—	—
Rwanda	64	58	75	66	64	68	1.7	3.0	18	1,196	2,440	2,701	21	4	66
Seychelles	—	—	—	74	71	76	—	—	58	66,742	28,050	4	4	0	82
Somalia	7	1	2	55	53	57	<0.1	<0.1	47	1,600	1,300	17,731	110	10	10
South Sudan	—	*5*[h]	—	55	53	56	1.5	2.6	21	456	—	4,583	41	5	9
Tanzania	38	32	53	66	64	68	3.2	5.7	36	485	2,920	20,015	33	6	7
Uganda	50	43	60	63	60	65	3.8	6.6	26	685	2,360	10,260	22	4	24
Zambia	50	48	69	62	59	65	7.7	13.8	45	527	3,300	11,777	63	9	23
Zimbabwe	*67*	*66*	*85*	59	56	62	8.7	14.4	32	408	2,380	9,316	62	7	31

MIDDLE AFRICA

	Population (millions) mid-2022	Births per 1,000 Population	Deaths per 1,000 Population	Rate of Natural Increase (%)	Net Migration Rate	Population (millions) mid-2035	Population (millions) mid-2050	Infant Mortality Rate[a]	Total Fertility Rate[b]	Percent of Population Ages <15	Percent of Population Ages 65+	Family Planning Among Married Women Ages 15-49[c]: Using All Methods (%)	Using Modern Methods (%)	Demand Satisfied by Modern Methods (%)[d]	Life Expectancy at Birth (Years): Total	Males	Females	Population Ages 15-49 With HIV/AIDS (%): Males	Females	Urban Population (%)	Population per km² of Arable Land	GNI per Capita, PPP[e]	SPECIAL FOCUS COVID-19 INDICATORS — Excess Deaths Due to COVID-19 Pandemic, 2020-2021 Annual Average[f]: No. of Excess Deaths	Excess Deaths per 100,000 Population	Excess Deaths as % of Total Deaths	Population Fully Vaccinated Against COVID-19 (%)[g]
AFRICA	1,419	33	9	2.4	-0	1,891	2,478	47	4.3	40	3	36	32	51	63	61	64	1.8	3.4	44	894	5,212	820,657	61	8	20
SUB-SAHARAN AFRICA	1,168	35	9	2.6	-0	1,591	2,122	50	4.6	42	3	33	29	47	60	58	62	2.2	4.2	42	739	3,993	608,098	55	7	17
MIDDLE AFRICA	196	40	9	3.1	-1	285	406	52	5.6	45	3	22	15	29	59	57	62	1.0	2.0	51	665	2,814	104,651	58	7	—
Angola	35.6	39	8	3.1	1	51.3	72.3	48	5.3	45	3	14	13	24	62	59	65	1.0	2.1	67	726	6,060	11,343	34	5	21
Cameroon	27.9	35	9	2.6	-0	38.1	51.3	48	4.5	42	3	19	15	36	60	59	62	1.8	3.9	58	450	3,990	17,661	66	8	5
Central African Republic	5.6	43	11	3.1	-16	8.2	11.5	71	6.0	48	3	18	14	26	54	52	56	2.1	3.4	43	310	1,090	3,123	64	5	22
Chad	17.7	43	12	3.1	-0	25.7	36.5	65	6.2	48	2	8	7	17	53	51	55	0.8	1.3	24	341	1,540	9,625	58	6	13
Congo	6.0	31	7	2.4	-1	7.9	10.4	32	4.2	41	3	*30*	*19*	*39*	64	62	65	2.1	5.6	68	1,086	2,850	2,265	41	6	—
Congo, Dem. Rep.	99.0	42	10	3.2	-1	148.0	217.5	52	6.2	47	3	28	18	31	59	57	62	0.6	0.9	46	735	1,170	58,879	65	8	—
Equatorial Guinea	1.7	30	9	2.1	4	2.2	2.8	57	4.3	38	3	*13*	*10*	*21*	61	59	63	5.2	9.4	74	1,208	13,520	860	60	8	15
Gabon	2.4	27	7	2.0	1	3.0	3.8	29	3.5	36	4	*31*	*19*	*34*	66	63	68	1.5	4.6	90	735	14,560	801	36	5	12
São Tomé and Príncipe	0.2	28	6	2.2	-4	0.3	0.4	12	3.8	39	4	50	46	60	68	65	70	0.4	0.5	75	5,685	4,490	96	43	9	42

SOUTHERN AFRICA

	Population (millions) mid-2022	Births per 1,000 Population	Deaths per 1,000 Population	Rate of Natural Increase (%)	Net Migration Rate	Population (millions)		Infant Mortality Rate[a]	Total Fertility Rate[b]	Percent of Population		Family Planning Among Married Women Ages 15-49[c]			Life Expectancy at Birth (Years)			Population Ages 15-49 With HIV/AIDS (%)		Urban Population (%)	Population per km² of Arable Land	GNI per Capita, PPP[e]	SPECIAL FOCUS COVID-19 INDICATORS: Excess Deaths Due to COVID-19 Pandemic, 2020-2021 Annual Average[f]			Population Fully Vaccinated Against COVID-19 (%)[g]
						mid-2035	mid-2050			Ages <15	Ages 65+	Using All Methods (%)	Using Modern Methods (%)	Demand Satisfied by Modern Methods (%)[d]	Total	Males	Females	Males	Females				No. of Excess Deaths	Excess Deaths per 100,000 Population	Excess Deaths as % of Total Deaths	
AFRICA	1,419	33	9	2.4	-0	1,891	2,478	47	4.3	40	3	36	32	51	63	61	64	1.8	3.4	44	894	5,212	820,657	61	8	20
SUB-SAHARAN AFRICA	1,168	35	9	2.6	-0	1,591	2,122	50	4.6	42	3	33	29	47	60	58	62	2.2	4.2	42	739	3,993	608,098	55	7	17
SOUTHERN AFRICA	69	21	12	0.9	2	81	92	28	2.4	29	6	56	55	78	61	59	64	12.3	24.4	65	517	13,602	130,820	192	18	33
Botswana	2.6	24	9	1.4	1	3.2	3.7	33	2.8	33	4	67	—	—	61	59	64	13.4	23.8	72	1,012	16,650	3,707	156	15	64
Eswatini	1.2	24	11	1.3	-4	1.4	1.7	38	2.8	35	4	*66*	*66*	*81*	57	53	61	19.4	36.1	24	687	8,870	1,910	164	14	29
Lesotho	2.3	26	14	1.2	-1	2.6	2.9	68	3.0	34	4	65	65	80	53	50	56	15.7	26.1	29	387	3,030	1,991	93	5	34
Namibia	2.6	27	11	1.7	-2	3.1	3.8	28	3.3	36	4	*56*	*55*	*75*	59	56	63	8.4	15.1	53	321	9,650	3,877	151	18	17
South Africa	60.6	20	12	0.8	3	70.2	80.3	25	2.4	28	6	55	54	78	62	59	64	12.1	24.5	68	505	14,140	119,336	200	18	32

NORTHERN AMERICA

	Population (millions) mid-2022	Births per 1,000 Population	Deaths per 1,000 Population	Rate of Natural Increase (%)	Net Migration Rate	Population (millions)		Infant Mortality Rate[a]	Total Fertility Rate[b]	Percent of Population		Family Planning Among Married Women Ages 15-49[c]			Life Expectancy at Birth (Years)			Population Ages 15-49 With HIV/AIDS (%)		Urban Population (%)	Population per km² of Arable Land	GNI per Capita, PPP[e]	SPECIAL FOCUS COVID-19 INDICATORS: Excess Deaths Due to COVID-19 Pandemic, 2020-2021 Annual Average[f]			Population Fully Vaccinated Against COVID-19 (%)[g]
						mid-2035	mid-2050			Ages <15	Ages 65+	Using All Methods (%)	Using Modern Methods (%)	Demand Satisfied by Modern Methods (%)[d]	Total	Males	Females	Males	Females				No. of Excess Deaths	Excess Deaths per 100,000 Population	Excess Deaths as % of Total Deaths	
AMERICAS	1,028	13	9	0.4	0	1,118	1,170	12	1.8	21	12	75	69	—	74	71	78	—	—	82	560	36,225	1,614,049	158	18	70
NORTHERN AMERICA	372	11	10	0.1	3	404	424	5	1.6	18	17	75	66	82	77	74	80	—	—	83	203	68,545	477,239	129	12	69
Canada	38.8	10	8	0.2	10	43.8	48.8	4.5	1.4	16	19	85[b]	—	—	82	80	84	—	—	82	101	51,690	11,010	29	4	82
United States of America	332.8	11	10	0.1	2	360.0	375.4	5.4	1.7	18	17	74	66	82	76	73	79	—	—	83	211	70,480	466,229	140	13	67

CENTRAL AMERICA

	Population (millions) mid-2022	Births per 1,000 Population	Deaths per 1,000 Population	Rate of Natural Increase (%)	Net Migration Rate	Population (millions)		Infant Mortality Rate[a]	Total Fertility Rate[b]	Percent of Population		Family Planning Among Married Women Ages 15-49[c]			Life Expectancy at Birth (Years)			Population Ages 15-49 With HIV/AIDS (%)		Urban Population (%)	Population per km² of Arable Land	GNI per Capita, PPP[e]	SPECIAL FOCUS COVID-19 INDICATORS: Excess Deaths Due to COVID-19 Pandemic, 2020-2021 Annual Average[f]			Population Fully Vaccinated Against COVID-19 (%)[g]
						mid-2035	mid-2050			Ages <15	Ages 65+	Using All Methods (%)	Using Modern Methods (%)	Demand Satisfied by Modern Methods (%)[d]	Total	Males	Females	Males	Females				No. of Excess Deaths	Excess Deaths per 100,000 Population	Excess Deaths as % of Total Deaths	
AMERICAS	1,028	13	9	0.4	0	1,118	1,170	12	1.8	21	12	75	69	—	74	71	78	—	—	82	560	36,225	1,614,049	158	18	70
LATIN AMERICA & THE CARIBBEAN	656	15	8	0.7	-1	715	746	14	1.9	23	10	75	70	—	73	70	77	0.7	0.3	81	764	16,633	1,136,810	175	22	70
CENTRAL AMERICA	178	16	8	0.7	-0	197	210	13	1.9	26	8	71	67	—	71	67	75	0.5	0.2	75	844	17,139	372,697	206	26	61
Belize	0.4	17	5	1.2	1	0.5	0.6	12	1.9	36	4	*51*	*49*	*66*	70	67	74	1.1	1.1	46	491	6,600	349	87	14	53
Costa Rica	5.2	11	6	0.4	6	5.7	6.1	8.5	1.3	22	9	71	69	83	81	78	83	0.8	0.1	81	2,133	21,800	4,824	94	16	81
El Salvador	6.3	16	9	0.7	-4	6.6	6.6	11	1.8	25	8	*72*	*68*	*82*	71	66	75	0.5	0.4	74	879	9,120	8,518	131	17	67
Guatemala	17.8	21	7	1.5	-2	21.3	24.6	19	2.4	32	5	*61*	*49*	*66*	69	66	73	0.3	0.2	52	2,070	9,580	24,698	137	20	35
Honduras	9.6	21	4	1.6	-1	11.7	13.4	15	2.4	31	6	*73*	*64*	*76*	70	68	73	0.3	0.2	59	961	5,740	11,331	113	17	54
Mexico	127.5	15	9	0.5	-0	138.2	143.8	11	1.8	25	8	73	70	—	70	66	75	0.6	0.2	81	635	19,540	313,109	242	29	63
Nicaragua	6.7	19	5	1.4	-3	7.9	8.8	13	2.2	30	6	*80*	*77*	*90*	77	74	80	0.4	0.2	59	448	5,960	6,048	91	15	80
Panama	4.4	15	5	1.0	4	5.2	5.9	12	2.0	25	10	*51*	*47*	*63*	77	74	80	—	—	69	778	29,900	3,822	88	15	72

CARIBBEAN

	Population (millions) mid-2022	Births per 1,000 Population	Deaths per 1,000 Population	Rate of Natural Increase (%)	Net Migration Rate	Population (millions) mid-2035	Population (millions) mid-2050	Infant Mortality Rate[a]	Total Fertility Rate[b]	Percent of Population Ages <15	Percent of Population Ages 65+
AMERICAS	**1,028**	**13**	**9**	**0.4**	**0**	**1,118**	**1,170**	**12**	**1.8**	**21**	**12**
LATIN AMERICA & THE CARIBBEAN	**656**	**15**	**8**	**0.7**	**-1**	**715**	**746**	**14**	**1.9**	**23**	**10**
CARIBBEAN	**44**	**15**	**10**	**0.5**	**-1**	**46**	**47**	**29**	**2.0**	**23**	**10**
Antigua and Barbuda	0.1	12	6	0.6	9	0.1	0.1	*6*	1.6	21	9
Bahamas	0.4	11	9	0.2	1	0.4	0.5	10	1.4	19	9
Barbados	0.3	8	9	-0.1	-0	0.3	0.3	11	1.3	17	16
Cuba	11.1	9	15	-0.6	0	10.7	9.9	7.6	1.5	16	16
Curaçao	0.1	9	11	-0.2	-14	0.2	0.2	*5.7*	1.4	17	20
Dominica	0.07	13	9	0.4	1	0.08	0.07	*12*	1.6	20	10
Dominican Republic	11.2	18	7	1.2	-1	12.4	13.2	27	2.3	27	7
Grenada	0.1	16	8	0.8	-1	0.1	0.1	*13*	2.0	24	10
Guadeloupe	0.4	11	14	-0.3	-2	0.4	0.4	*7.5*	2.1	18	21
Haiti	11.6	23	9	1.5	-3	13.4	15.1	46	2.8	32	5
Jamaica	2.8	12	9	0.3	-2	2.8	2.5	11	1.4	20	7
Martinique	0.3	10	14	-0.4	-4	0.3	0.3	*7*	1.9	15	23
Puerto Rico	3.2	6	10	-0.4	-1	3.1	2.7	7.3	0.9	13	23
St. Kitts-Nevis	0.05	12	10	0.2	-2	0.05	0.05	*8*	1.5	20	10
St. Lucia	0.2	11	10	0.1	0	0.2	0.2	*12*	1.4	18	9
St. Vincent and the Grenadines	0.1	12	9	0.3	-1	0.1	0.1	*14*	1.6	25	9
Trinidad and Tobago	1.4	11	8	0.3	-1	1.4	1.3	16	1.6	21	9

	Family Planning Among Married Women Ages 15-49[c]: Using All Methods (%)	Using Modern Methods (%)	Demand Satisfied by Modern Methods (%)[d]	Life Expectancy at Birth (Years): Total	Males	Females	Population Ages 15-49 With HIV/AIDS (%): Males	Females	Urban Population (%)	Population per km² of Arable Land	GNI per Capita, PPP[e]	SPECIAL FOCUS COVID-19 INDICATORS — Excess Deaths Due to COVID-19 Pandemic, 2020-2021 Annual Average[f]: No. of Excess Deaths	Excess Deaths per 100,000 Population	Excess Deaths as % of Total Deaths	Population Fully Vaccinated Against COVID-19 (%)[g]
AMERICAS	**75**	**69**	**—**	**74**	**71**	**78**	**—**	**—**	**82**	**560**	**36,225**	**1,614,049**	**158**	**18**	**70**
LATIN AMERICA & THE CARIBBEAN	**75**	**70**	**—**	**73**	**70**	**77**	**0.7**	**0.3**	**81**	**764**	**16,633**	**1,136,810**	**175**	**22**	**70**
CARIBBEAN	**57**	**55**	**71**	**71**	**68**	**74**	**1.1**	**1.2**	**72**	**1,793**	**—**	**23,144**	**59**	**7**	**46**
Antigua and Barbuda	—	—	—	79	76	81	—	—	24	2,524	19,610	-15	-15	-2	63
Bahamas	—	—	—	72	68	75	1.4	1.2	83	5,125	31,870	502	127	15	40
Barbados	*59*	*55*	*70*	78	76	79	1.1	0.6	31	3,846	14,530	-178	-62	-5	53
Cuba	69	68	87	74	71	76	0.8	0.2	77	381	—	9,080	80	7	88
Curaçao	—	—	—	78	75	82	—	—	89	—	*21,480*	—	—	—	—
Dominica	—	—	—	73	70	76	—	—	71	1,212	12,010	8	10	1	42
Dominican Republic	*70*	*68*	*84*	73	69	76	1.0	0.9	83	1,280	19,730	5,937	54	7	55
Grenada	—	—	—	75	72	78	—	—	37	4,181	15,890	-134	-118	-15	34
Guadeloupe	—	—	—	76	73	80	—	—	99	1,703	—	—	—	—	—
Haiti	34	32	44	63	60	66	1.4	2.2	58	1,083	3,130	4,805	42	5	1
Jamaica	*73*	*68*	*83*	71	68	73	1.3	1.4	57	2,356	9,720	1,796	61	8	24
Martinique	—	—	—	78	74	81	—	—	89	3,177	—	—	—	—	—
Puerto Rico	—	—	—	80	76	85	—	—	94	6,466	*23,360*	—	—	—	—
St. Kitts-Nevis	—	—	—	72	68	75	—	—	31	953	25,900	-104	-194	-25	49
St. Lucia	*56*	*53*	*72*	71	68	75	—	—	19	6,736	13,810	185	101	10	29
St. Vincent and the Grenadines	—	—	—	71	68	75	—	—	53	5,521	13,950	248	222	21	28
Trinidad and Tobago	*40*	*38*	*58*	76	74	78	0.9	1.2	53	5,476	25,670	1,016	72	10	51

SOUTH AMERICA

	Population (millions) mid-2022	Births per 1,000 Population	Deaths per 1,000 Population	Rate of Natural Increase (%)	Net Migration Rate	Population (millions)		Infant Mortality Rate[a]	Total Fertility Rate[b]	Percent of Population		Family Planning Among Married Women Ages 15-49[c]			Life Expectancy at Birth (Years)			Population Ages 15-49 With HIV/AIDS (%)		Urban Population (%)	Population per km² of Arable Land	GNI per Capita, PPP[e]	SPECIAL FOCUS COVID-19 INDICATORS: Excess Deaths Due to COVID-19 Pandemic, 2020-2021 Annual Average[f]			Population Fully Vaccinated Against COVID-19 (%)[g]
						mid-2035	mid-2050			Ages <15	Ages 65+	Using All Methods (%)	Using Modern Methods (%)	Demand Satisfied by Modern Methods (%)[d]	Total	Males	Females	Males	Females				No. of Excess Deaths	Excess Deaths per 100,000 Population	Excess Deaths as % of Total Deaths	
AMERICAS	1,028	13	9	0.4	0	1,118	1,170	12	1.8	21	12	75	69	—	74	71	78	—	—	82	560	36,225	1,614,049	158	18	70
LATIN AMERICA & THE CARIBBEAN	656	15	8	0.7	-1	715	746	14	1.9	23	10	75	70	—	73	70	77	0.7	0.3	81	764	16,633	1,136,810	175	22	70
SOUTH AMERICA	434	14	8	0.6	-1	471	488	14	1.8	22	10	78	73	—	74	71	78	0.7	0.3	85	628	16,521	740,970	172	21	76
Argentina	46.2	14	9	0.5	0	51.2	54.0	9.3	1.9	23	12	70	68	81	75	72	79	0.5	0.3	92	142	23,150	44,941	99	11	83
Bolivia	12.2	22	10	1.2	-0	14.5	16.6	27	2.6	31	5	67	45	50	64	61	67	0.4	0.2	70	269	8,800	44,015	375	33	51
Brazil	214.8	13	7	0.6	-0	229.2	232.9	14	1.6	20	10	80	78	—	75	72	79	0.7	0.4	87	385	15,550	340,634	160	19	79
Chile	19.8	11	6	0.5	12	21.1	21.6	6.5	1.4	19	13	76	70	—	81	78	84	1.1	0.2	88	1,657	27,410	19,348	101	15	92
Colombia	49.1	15	8	0.8	-1	51.5	51.5	12	2.0	23	11	81	76	86	75	71	79	0.9	0.2	82	1,006	16,460	82,372	161	25	71
Ecuador	18.0	17	7	1.0	2	20.3	22.3	11	2.0	26	8	80	72	81	74	70	77	0.5	0.3	64	1,734	11,480	40,434	228	34	79
French Guiana	0.3	27	5	2.2	-4	0.4	0.5	8	3.5	32	6	—	—	—	76	73	80	—	—	86	2,232	—	—	—	—	—
Guyana	0.8	20	10	1.0	-5	0.9	0.9	23	2.4	29	6	34	33	53	66	62	69	1.3	1.5	27	193	23,480	1,406	178	14	47
Paraguay	6.8	21	8	1.3	-1	7.8	8.6	15	2.5	29	6	68	67	83	70	67	73	0.4	0.3	62	143	13,630	9,909	138	22	48
Peru	33.4	17	12	0.6	-1	36.4	39.4	15	2.2	25	9	76	55	67	69	65	73	0.5	0.2	79	934	12,900	144,834	437	49	83
Suriname	0.6	17	7	1.0	-3	0.7	0.8	17	2.2	26	8	39	39	57	73	71	76	1.4	1.3	66	986	14,430	364	62	7	41
Uruguay	3.6	10	12	-0.2	-0	3.6	3.5	4.7	1.5	19	16	80[h]	78[h]	—	75	72	79	0.7	0.5	96	176	22,540	1,553	45	4	83
Venezuela	28.3	16	8	0.8	-18	33.4	35.9	16	2.2	28	9	75	—	—	71	66	75	0.7	0.3	88	1,089	—	11,163	39	6	50

WESTERN ASIA

2022 WORLD POPULATION DATA SHEET | 12

	Population (millions) mid-2022	Births per 1,000 Population	Deaths per 1,000 Population	Rate of Natural Increase (%)	Net Migration Rate	Population (millions) mid-2035	Population (millions) mid-2050	Infant Mortality Rate[a]	Total Fertility Rate[b]	Percent of Population Ages <15	Percent of Population Ages 65+	Family Planning Among Married Women Ages 15-49[c]: Using All Methods (%)	Using Modern Methods (%)	Demand Satisfied by Modern Methods (%)[d]
ASIA	4,730	15	7	0.8	-0	5,094	5,313	23	1.9	24	10	67	59	—
WESTERN ASIA	**294**	**19**	**5**	**1.3**	**-1**	**355**	**411**	**18**	**2.5**	**28**	**6**	**52**	**37**	**55**
Armenia	3.0	12	12	0.1	1	2.8	2.5	7	1.7	20	12	57	28	40
Azerbaijan	10.2	13	8	0.5	0	10.9	11.2	10	1.7	22	7	*55*	*14*	—
Bahrain	1.5	12	2	0.9	-6	1.6	1.8	6	1.8	20	4	—	—	—
Cyprus	1.3	10	7	0.3	2	1.3	1.4	2	1.3	16	15	—	—	—
Georgia	3.7	12	16	-0.4	-1	3.6	3.4	9	2.0	20	15	41	33	51
Iraq	44.5	27	5	2.3	-1	58.3	74.5	20	3.5	38	3	53	36	54
Israel	9.5	20	5	1.4	2	12.1	15.7	2.8	3.0	28	12	—	—	—
Jordan	11.2	18	6	1.2	1	13.0	14.1	17	2.4	34	4	52	37	57
Kuwait	4.1	12	2	0.9	-9	4.6	5.0	7	2.0	21	4	—	—	—
Lebanon	5.5	15	8	0.7	-20	4.6	4.9	6	2.1	24	11	*55*	*47*	—
Oman	4.6	19	2	1.6	-6	5.4	6.3	8	2.7	27	3	*30*	*19*	*40*
Palestinian Territory	5.4	29	4	2.5	-2	7.0	9.0	12	3.7	39	4	57	43	61
Qatar	2.7	10	1	0.9	-11	3.0	3.4	5	1.8	16	2	*38*	*34*	69
Saudi Arabia	36.7	17	3	1.5	-4	43.0	48.7	6	2.4	24	3	25	21	—
Syria	22.1	20	5	1.5	10	32.8	38.3	18	2.7	31	5	*54*	*38*	*53*
Türkiye	85.2	13	6	0.6	-1	96.4	103.9	8	1.7	22	10	70	49	60
United Arab Emirates	9.4	10	2	0.8	-0	10.3	11.4	5	1.5	15	2	—	—	—
Yemen	33.7	30	7	2.3	-1	43.9	55.3	47	3.7	39	3	*34*	*29*	*47*

	Life Expectancy at Birth (Years): Total	Males	Females	Population Ages 15-49 With HIV/AIDS (%): Males	Females	Urban Population (%)	Population per km² of Arable Land	GNI per Capita, PPP[e]	SPECIAL FOCUS COVID-19 INDICATORS — Excess Deaths Due to COVID-19 Pandemic, 2020-2021 Annual Average[f]: No. of Excess Deaths	Excess Deaths per 100,000 Population	Excess Deaths as % of Total Deaths	Population Fully Vaccinated Against COVID-19 (%)[g]
ASIA	73	71	76	—	—	52	3,168	14,638	3,663,240	80	10	71
WESTERN ASIA	**74**	**72**	**77**	**—**	**—**	**72**	**3,354**	**29,448**	**274,239**	**97**	**15**	**46**
Armenia	74	68	79	0.4	0.2	64	667	14,220	9,832	332	28	33
Azerbaijan	73	70	77	0.2	0.1	53	489	15,460	28,504	280	26	48
Bahrain	79	78	80	0.2	<0.1	90	92,015	*40,730*	333	19	6	72
Cyprus	81	79	83	0.2	<0.1	67	1,222	39,410	505	42	7	53
Georgia	73	69	78	0.4	0.2	59	1,208	16,060	12,233	307	19	32
Iraq	70	68	72	—	—	71	890	10,320	33,368	82	17	19
Israel	83	81	85	—	—	93	2,507	44,060	3,089	35	6	71
Jordan	73	72	75	—	—	90	5,429	10,900	5,965	58	17	44
Kuwait	79	77	80	—	—	100	51,868	*59,040*	2,120	49	19	78
Lebanon	75	73	77	0.1	<0.1	89	4,057	10,360	9,260	136	19	—
Oman	73	71	75	0.2	<0.1	87	6,127	*28,850*	5,750	111	26	60
Palestinian Territory	73	71	76	—	—	77	7,672	7,370	—	—	—	—
Qatar	79	78	81	0.2	<0.1	99	12,834	92,080	770	26	14	92
Saudi Arabia	77	76	79	—	—	85	1,069	*47,700*	8,775	25	6	72
Syria	72	69	75	<0.1	<0.1	56	475	—	3,657	20	5	10
Türkiye	79	76	81	—	—	77	435	30,020	132,021	156	23	63
United Arab Emirates	80	78	81	—	—	87	19,268	*66,680*	1,165	12	5	98
Yemen	64	60	67	<0.1	<0.1	39	2,910	—	16,896	56	9	1

CENTRAL ASIA

	Population (millions) mid-2022	Births per 1,000 Population	Deaths per 1,000 Population	Rate of Natural Increase (%)	Net Migration Rate	Population (millions)		Infant Mortality Rate[a]	Total Fertility Rate[b]	Percent of Population		Family Planning Among Married Women Ages 15-49[c]			Life Expectancy at Birth (Years)			Population Ages 15-49 With HIV/AIDS (%)		Urban Population (%)	Population per km² of Arable Land	GNI per Capita, PPP[e]	SPECIAL FOCUS COVID-19 INDICATORS: Excess Deaths Due to COVID-19 Pandemic, 2020-2021 Annual Average[f]			Population Fully Vaccinated Against COVID-19 (%)[g]
						mid-2035	mid-2050			Ages <15	Ages 65+	Using All Methods (%)	Using Modern Methods (%)	Demand Satisfied by Modern Methods (%)[d]	Total	Males	Females	Males	Females				No. of Excess Deaths	Excess Deaths per 100,000 Population	Excess Deaths as % of Total Deaths	
ASIA	4,730	15	7	0.8	-0	5,094	5,313	23	1.9	24	10	67	59	—	73	71	76	—	—	52	3,168	14,638	3,663,240	80	10	71
CENTRAL ASIA	**78**	**25**	**6**	**1.9**	**-0**	**91**	**105**	**14**	**3.0**	**31**	**5**	**54**	**51**	**—**	**72**	**69**	**75**	**0.3**	**0.2**	**48**	**641**	**12,505**	**73,538**	**98**	**14**	**45**
Kazakhstan	19.2	23	10	1.4	-0	22.0	25.4	8	3.3	30	8	53	50	73	70	66	74	0.4	0.2	58	65	24,970	38,108	202	22	50
Kyrgyzstan	6.8	24	6	1.8	-1	8.1	9.7	14	3.0	34	4	39	38	65	72	68	76	0.3	0.2	37	528	4,840	6,157	94	16	20
Tajikistan	10.0	27	5	2.2	-0	12.4	15.2	27	3.2	36	3	29	27	52	72	70	74	0.2	0.2	28	1,186	5,180	6,505	67	12	52
Turkmenistan	6.4	22	7	1.5	-1	7.4	8.3	35	2.7	31	5	50	47	80	69	66	73	—	—	53	331	*15,600*	326	5	1	—
Uzbekistan	35.6	26	5	2.1	-1	41.4	46.9	9	2.9	29	5	65	62	—	73	71	76	0.3	0.2	50	886	8,520	22,443	67	12	46

SOUTH ASIA

	Population (millions) mid-2022	Births per 1,000 Population	Deaths per 1,000 Population	Rate of Natural Increase (%)	Net Migration Rate	Population (millions)		Infant Mortality Rate[a]	Total Fertility Rate[b]	Percent of Population		Family Planning Among Married Women Ages 15-49[c]			Life Expectancy at Birth (Years)			Population Ages 15-49 With HIV/AIDS (%)		Urban Population (%)	Population per km² of Arable Land	GNI per Capita, PPP[e]	SPECIAL FOCUS COVID-19 INDICATORS: Excess Deaths Due to COVID-19 Pandemic, 2020-2021 Annual Average[f]			Population Fully Vaccinated Against COVID-19 (%)[g]
						mid-2035	mid-2050			Ages <15	Ages 65+	Using All Methods (%)	Using Modern Methods (%)	Demand Satisfied by Modern Methods (%)[d]	Total	Males	Females	Males	Females				No. of Excess Deaths	Excess Deaths per 100,000 Population	Excess Deaths as % of Total Deaths	
ASIA	4,730	15	7	0.8	-0	5,094	5,313	23	1.9	24	10	67	59	—	73	71	76	—	—	52	3,168	14,638	3,663,240	80	10	71
SOUTH ASIA	**2,008**	**20**	**6**	**1.4**	**-1**	**2,268**	**2,477**	**31**	**2.2**	**27**	**6**	**62**	**52**	**74**	**70**	**68**	**71**	**0.2**	**0.2**	**37**	**996**	**6,958**	**2,706,158**	**138**	**17**	**63**
Afghanistan	41.1	36	7	2.8	-5	56.2	74.1	45	4.6	43	2	19[h]	18[h]	—	62	59	65	<0.1	<0.1	26	525	*2,100*	22,267	57	8	13
Bangladesh	171.2	18	6	1.2	-1	191.2	203.9	23	2.0	26	6	63	59	77	72	71	74	<0.1	<0.1	39	2,140	6,960	70,382	43	8	—
Bhutan	0.8	14	4	1.0	0	0.8	0.9	15	1.5	24	6	*66*	*65*	*85*	72	70	74	0.2	0.2	43	812	*10,440*	-199	-26	-4	87
India	1,417.2	20	6	1.4	-0	1,567.8	1,670.5	28	2.1	25	7	67	56	74	70	68	71	0.2	0.2	35	912	7,220	2,370,447	171	20	66
Iran	88.6	14	6	0.7	0	94.9	99.0	11	1.7	24	8	*77*	*57*	69	74	71	77	0.1	<0.1	76	566	*15,760*	116,077	137	22	69
Maldives	0.6	13	3	1.0	2	0.6	0.6	7	1.7	20	3	19	15	30	80	79	81	—	—	41	14,855	16,520	123	23	8	71
Nepal	30.9	20	8	1.3	10	34.8	37.8	23	2.0	29	6	47	44	62	68	67	70	0.1	0.1	21	1,462	4,280	16,257	55	8	—
Pakistan	235.8	28	7	2.0	-2	298.4	367.8	52	3.5	37	4	34	23	—	66	64	69	0.3	<0.1	37	762	5,800	115,220	52	7	57
Sri Lanka	22.4	13	7	0.6	-4	22.9	22.4	6	1.7	25	8	65	54	74	76	73	80	<0.1	<0.1	19	1,632	13,790	-4,416	-21	-3	68

SOUTHEAST ASIA

	Population (millions) mid-2022	Births per 1,000 Population	Deaths per 1,000 Population	Rate of Natural Increase (%)	Net Migration Rate	Population (millions) mid-2035	Population (millions) mid-2050	Infant Mortality Rate[a]	Total Fertility Rate[b]	Percent of Population Ages <15	Percent of Population Ages 65+	Family Planning Among Married Women Ages 15-49[c]: Using All Methods (%)	Using Modern Methods (%)	Demand Satisfied by Modern Methods (%)[d]	Life Expectancy at Birth (Years): Total	Males	Females	Population Ages 15-49 With HIV/AIDS (%): Males	Females	Urban Population (%)	Population per km² of Arable Land	GNI per Capita, PPP[e]	SPECIAL FOCUS COVID-19 INDICATORS — Excess Deaths Due to COVID-19 Pandemic, 2020-2021 Annual Average[f]: No. of Excess Deaths	Excess Deaths per 100,000 Population	Excess Deaths as % of Total Deaths	Population Fully Vaccinated Against COVID-19 (%)[g]
ASIA	4,730	15	7	0.8	-0	5,094	5,313	23	1.9	24	10	67	59	—	73	71	76	—	—	52	3,168	14,638	3,663,240	80	10	71
SOUTHEAST ASIA	**676**	**17**	**8**	**0.8**	**0**	**742**	**786**	**18**	**2.2**	**25**	**8**	**63**	**54**	**73**	**70**	**68**	**73**	**0.5**	**0.3**	**51**	**9,885**	**13,147**	**645,476**	**96**	**11**	**67**
Brunei	0.4	14	4	1.0	-2	0.5	0.5	8	1.8	20	6	—	—	—	79	78	81	—	—	79	11,051	*67,580*	-55	-13	-2	98
Cambodia	16.8	22	7	1.5	-2	19.5	22.2	12	2.7	29	6	62	45	61	70	67	73	0.5	0.6	25	433	4,430	6,259	37	6	86
Indonesia	275.5	16	10	0.7	-0	300.9	317.2	18	2.2	25	7	64	57	77	68	66	70	0.4	0.3	57	1,048	12,560	514,283	187	22	62
Laos	7.5	21	7	1.4	1	8.7	9.7	49	2.5	32	4	54	49	72	67	65	69	0.4	0.3	37	609	8,150	878	12	2	70
Malaysia	32.7	14	5	0.9	7	36.7	39.6	6	1.7	24	7	*52*	*34*	—	76	73	78	0.5	0.1	75	3,962	28,730	3,767	12	2	84
Myanmar	54.2	17	10	0.7	-1	58.2	59.9	34	2.2	25	7	52	51	75	66	63	69	1.0	0.7	31	492	4,230	22,094	40	6	50
Philippines	115.6	22	7	1.5	-1	137.6	157.9	21	2.7	30	5	54	40	57	69	67	71	0.4	<0.1	48	2,067	9,450	92,627	84	11	64
Singapore	5.5	9	6	0.3	3	5.8	5.8	1.8	1.1	14	16	—	—	—	84	82	86	0.4	<0.1	100	934,086	102,450	738	13	3	85
Thailand	66.8	10	8	0.2	0	66.8	63.2	6	1.5	17	12	73	71	88	77	74	81	1.2	0.9	59	397	18,530	7,651	11	1	76
Timor-Leste	1.3	25	7	1.8	-2	1.6	1.8	33	3.1	35	5	26	24	47	68	66	70	0.2	0.1	32	1,203	5,360	342	26	4	—
Vietnam	99.4	15	7	0.8	-0	105.9	108.4	10	2.2	22	9	73	60	72	74	69	78	0.4	0.2	37	1,465	11,040	-3,106	-3	0	82

EAST ASIA

	Population (millions) mid-2022	Births per 1,000 Population	Deaths per 1,000 Population	Rate of Natural Increase (%)	Net Migration Rate	Population (millions)		Infant Mortality Rate[a]	Total Fertility Rate[b]	Percent of Population		Family Planning Among Married Women Ages 15-49[c]			Life Expectancy at Birth (Years)			Population Ages 15-49 With HIV/AIDS (%)		Urban Population (%)	Population per km² of Arable Land	GNI per Capita, PPP[e]	SPECIAL FOCUS COVID-19 INDICATORS: Excess Deaths Due to COVID-19 Pandemic, 2020-2021 Annual Average[f]			Population Fully Vaccinated Against COVID-19 (%)[g]
						mid-2035	mid-2050			Ages <15	Ages 65+	Using All Methods (%)	Using Modern Methods (%)	Demand Satisfied by Modern Methods (%)[d]	Total	Males	Females	Males	Females				No. of Excess Deaths	Excess Deaths per 100,000 Population	Excess Deaths as % of Total Deaths	
ASIA	4,730	15	7	0.8	-0	5,094	5,313	23	1.9	24	10	67	59	—	73	71	76	—	—	52	3,168	14,638	3,663,240	80	10	71
EAST ASIA	**1,674**	**7**	**8**	**0.0**	**-0**	**1,639**	**1,533**	**6**	**1.2**	**18**	**15**	**81**	**77**	**—**	**79**	**76**	**82**	**—**	**—**	**68**	**3,122**	**22,276**	**-36,170**	**-2**	**0**	**88**
China	1,436.6	8	7	0.0	-0	1,410.1	1,322.5	6	1.2	18	14	85	81	—	78	75	81	—	—	65	1,208	19,170	-26,032	-2	0	89
China, Hong Kong SAR[i]	7.4	5	7	-0.2	-1	7.8	7.7	1.9	0.8	11	20	67	65	—	85	83	88	—	—	100	369,566	70,700	—	—	—	—
China, Macao SAR[i]	0.7	7	3	0.4	6	0.8	0.9	*2*	0.8	15	12	—	—	—	84	81	87	—	—	100	—	*72,260*	—	—	—	—
Japan	124.9	7	12	-0.5	-1	115.2	101.8	1.7	1.3	12	29	*40*	*33*	—	85	82	88	—	—	92	3,044	44,570	-9,736	-8	-1	81
Korea, North	26.1	13	9	0.4	-0	26.6	25.8	14	1.8	19	12	70	69	90	74	71	76	—	—	63	1,143	—	-3,549	-14	-2	—
Korea, South	51.6	5	6	-0.1	0	50.9	47.4	2.5	0.8	12	16	82	—	—	84	81	87	—	—	81	3,819	47,490	3,144	6	1	—
Mongolia	3.4	22	6	1.5	0	4.2	5.2	6	2.8	32	4	48	45	64	71	67	76	<0.1	<0.1	69	256	11,090	2	0	0	66
Taiwan	23.2	7	8	-0.1	7	23.1	21.8	3.6	1.0	12	17	—	—	—	81	78	84	—	—	79	3,929	—	—	—	—	81

NORTHERN EUROPE

	Population (millions) mid-2022	Births per 1,000 Population	Deaths per 1,000 Population	Rate of Natural Increase (%)	Net Migration Rate	Population (millions)		Infant Mortality Rate[a]	Total Fertility Rate[b]	Percent of Population		Family Planning Among Married Women Ages 15-49[c]			Life Expectancy at Birth (Years)			Population Ages 15-49 With HIV/AIDS (%)		Urban Population (%)	Population per km^2 of Arable Land	GNI per Capita, PPP[e]	SPECIAL FOCUS COVID-19 INDICATORS: Excess Deaths Due to COVID-19 Pandemic, 2020-2021 Annual Average[f]			Population Fully Vaccinated Against COVID-19 (%)[g]
						mid-2035	mid-2050			Ages <15	Ages 65+	Using All Methods (%)	Using Modern Methods (%)	Demand Satisfied by Modern Methods (%)[d]	Total	Males	Females	Males	Females				No. of Excess Deaths	Excess Deaths per 100,000 Population	Excess Deaths as % of Total Deaths	
EUROPE	742	9	13	-0.3	2	740	724	4	1.5	16	19	66	56	—	78	75	81	—	—	75	555	42,820	1,365,595	182	13	68
EUROPEAN UNION	445	9	12	-0.3	3	445	439	3	1.5	15	21	65	—	—	80	78	83	—	—	75	613	48,451	566,819	128	10	74
NORTHERN EUROPE	107	10	10	0.0	3	110	113	3	1.6	18	19	70	66	—	81	79	83	—	—	83	928	54,447	98,916	93	9	75
Channel Islands	0.2	10	8	0.2	7	0.2	0.2	*7.3*	1.5	15	16	—	—	—	81	79	83	—	—	31	4,796	—	—	—	—	—
Denmark	5.9	11	10	0.1	5	6.1	6.3	3.1	1.7	16	20	—	—	—	82	80	83	0.1	<0.1	88	248	66,720	1,858	32	3	82
Estonia	1.3	9	14	-0.4	5	1.3	1.3	*2.2*	1.5	16	20	—	—	—	79	74	83	—	—	69	191	41,570	1,687	127	10	64
Finland	5.6	9	10	-0.1	4	5.6	5.5	1.8	1.5	15	23	*86*	*82*	—	82	79	84	—	—	87	248	55,940	1,429	26	2	78
Iceland	0.4	13	6	0.7	13	0.4	0.5	*3.3*	1.8	19	15	—	—	—	83	81	84	0.2	<0.1	94	314	55,920	-6	-2	0	85
Ireland	5.1	12	7	0.5	2	5.4	5.8	3.1	1.8	20	15	*73*	*71*	—	82	80	84	0.3	0.2	64	1,138	79,450	1,461	29	4	82
Latvia	1.9	9	18	-0.9	-2	1.6	1.4	2.7	1.6	16	21	—	—	—	75	70	80	0.9	0.5	80	140	33,790	3,834	204	11	69
Lithuania	2.8	9	17	-0.8	7	2.5	2.1	3.1	1.4	15	20	*69*	*52*	—	75	70	80	0.4	<0.1	68	124	41,250	8,628	319	17	69
Norway	5.4	10	8	0.3	4	5.8	6.0	1.9	1.6	17	18	—	—	—	83	82	85	0.1	<0.1	82	676	82,840	-51	-1	0	75
Sweden	10.5	11	9	0.2	4	11.2	11.8	1.8	1.7	18	20	—	—	—	83	81	85	—	—	88	413	61,090	5,628	56	6	76
United Kingdom	67.6	10	10	0.0	3	70.1	71.8	4.0	1.6	18	19	*76*	*71*	*84*	81	79	83	—	—	84	1,131	49,420	74,449	109	11	74

WESTERN EUROPE

	Population (millions) mid-2022	Births per 1,000 Population	Deaths per 1,000 Population	Rate of Natural Increase (%)	Net Migration Rate	Population (millions) mid-2035	Population (millions) mid-2050	Infant Mortality Rate[a]	Total Fertility Rate[b]	Percent of Population Ages <15	Percent of Population Ages 65+	Family Planning Among Married Women Ages 15-49[c]: Using All Methods (%)	Using Modern Methods (%)	Demand Satisfied by Modern Methods (%)[d]	Life Expectancy at Birth (Years): Total	Males	Females	Population Ages 15-49 With HIV/AIDS (%): Males	Females	Urban Population (%)	Population per km² of Arable Land	GNI per Capita, PPP[e]	SPECIAL FOCUS COVID-19 INDICATORS — Excess Deaths Due to COVID-19 Pandemic, 2020-2021 Annual Average[f]: No. of Excess Deaths	Excess Deaths per 100,000 Population	Excess Deaths as % of Total Deaths	Population Fully Vaccinated Against COVID-19 (%)[g]
EUROPE	742	9	13	-0.3	2	740	724	4	1.5	16	19	66	56	—	78	75	81	—	—	75	555	42,820	1,365,595	182	13	68
EUROPEAN UNION	445	9	12	-0.3	3	445	439	3	1.5	15	21	65	—	—	80	78	83	—	—	75	613	48,451	566,819	128	10	74
WESTERN EUROPE	197	10	11	-0.1	1	202	204	3	1.6	15	21	72	—	—	82	79	84	—	—	80	797	58,151	172,143	88	8	78
Austria	9.0	10	10	-0.1	6	9.4	9.6	2.7	1.5	14	19	79	—	—	81	79	84	—	—	59	681	58,370	5,969	66	6	74
Belgium	11.6	10	10	0.0	5	12.1	12.5	2.9	1.6	17	20	67	65	—	81	79	83	—	—	98	1,343	59,460	8,959	77	7	80
France	65.8	11	10	0.1	2	68.8	71.3	3.4	1.8	17	21	78	73	—	82	79	86	0.4	0.2	81	366	51,850	40,925	63	6	81
Germany	83.3	10	12	-0.3	-3	82.4	79.8	3.0	1.6	14	22	67[h]	—	—	81	79	83	—	—	78	714	59,680	97,494	116	10	78
Liechtenstein	0.04	9	8	0.1	7	0.04	0.04	1.9	1.5	15	19	—	—	—	82	80	84	—	—	14	2,798	—	—	—	—	69
Luxembourg	0.7	10	7	0.3	13	0.7	0.8	4.7	1.4	16	15	—	—	—	83	80	85	0.3	0.2	92	1,047	83,230	35	6	1	74
Monaco	0.04	25	16	1.0	11	0.04	0.04	4	2.1	13	36	—	—	—	86	84	88	—	—	100	—	—	32	81	6	65
Netherlands	17.7	10	9	0.1	7	18.8	19.5	3.8	1.6	15	20	73	70	—	81	80	83	0.2	<0.1	93	1,759	63,360	14,606	85	9	70
Switzerland	8.8	10	8	0.2	6	9.7	10.4	3.6	1.5	15	19	72	67	—	84	82	86	—	—	74	2,193	75,860	4,124	47	6	69

EASTERN EUROPE

	Population (millions) mid-2022	Births per 1,000 Population	Deaths per 1,000 Population	Rate of Natural Increase (%)	Net Migration Rate	Population (millions) mid-2035	Population (millions) mid-2050	Infant Mortality Rate[a]	Total Fertility Rate[b]	Percent of Population Ages <15	Percent of Population Ages 65+	Family Planning Among Married Women Ages 15-49[c]: Using All Methods (%)	Using Modern Methods (%)	Demand Satisfied by Modern Methods (%)[d]	Life Expectancy at Birth (Years): Total	Males	Females	Population Ages 15-49 With HIV/AIDS (%): Males	Females	Urban Population (%)	Population per km² of Arable Land	GNI per Capita, PPP[e]	SPECIAL FOCUS COVID-19 INDICATORS — Excess Deaths Due to COVID-19 Pandemic, 2020-2021 Annual Average[f]: No. of Excess Deaths	Excess Deaths per 100,000 Population	Excess Deaths as % of Total Deaths	Population Fully Vaccinated Against COVID-19 (%)[g]
EUROPE	742	9	13	-0.3	2	740	724	4	1.5	16	19	66	56	—	78	75	81	—	—	75	555	42,820	1,365,595	182	13	68
EUROPEAN UNION	445	9	12	-0.3	3	445	439	3	1.5	15	21	65	—	—	80	78	83	—	—	75	613	48,451	566,819	128	10	74
EASTERN EUROPE	287	9	15	-0.6	3	280	264	5	1.4	17	17	62	50	—	73	68	77	—	—	70	182	30,062	878,698	300	20	52
Belarus	9.2	9	13	-0.4	1	8.7	8.1	2.5	1.4	17	16	53	46	66	74	69	79	0.5	0.4	78	163	20,870	24,447	259	17	66
Bulgaria	6.8	9	22	-1.3	2	6.3	5.8	5.6	1.6	15	22	*69*	*40*	—	74	70	77	0.1	<0.1	73	195	26,000	28,747	415	21	30
Czechia	10.5	11	13	-0.3	5	10.8	10.8	2.2	1.8	16	21	*86*	*78*	*86*	77	74	81	—	—	74	424	42,560	18,520	173	14	64
Hungary	9.7	9	15	-0.6	0	9.5	9.2	3.3	1.5	15	20	*62*	*54*	—	76	72	79	—	—	72	241	35,640	18,249	189	12	64
Moldova	3.5	11	17	-0.6	-3	3.4	3.2	9	1.7	16	12	56	44	—	71	67	75	1.1	0.6	43	207	16,070	9,079	225	19	26
Poland	38.0	9	14	-0.5	14	36.5	34.0	3.6	1.3	15	19	*62*	—	—	77	73	81	—	—	60	348	35,830	78,766	208	16	59
Romania	19.0	9	17	-0.8	1	18.1	16.9	5.6	1.5	16	19	—	—	—	71	67	75	0.2	0.1	54	213	34,820	53,456	279	17	42
Russia	144.3	10	15	-0.5	1	143.1	136.8	4.5	1.5	18	16	*68*	*55*	*72*	72	66	76	—	—	75	119	32,000	536,163	367	23	51
Slovakia	5.4	10	14	-0.3	0	5.4	5.1	5.6	1.6	16	17	—	—	—	77	73	80	<0.1	<0.1	54	403	32,450	12,160	223	18	51
Ukraine	41.0	7	17	-1.1	1	38.1	33.9	6.6	1.0	15	17	65	*48*	*68*	71	66	76	1.0	0.9	70	124	13,860	99,113	227	15	—

SOUTHERN EUROPE

	Population (millions) mid-2022	Births per 1,000 Population	Deaths per 1,000 Population	Rate of Natural Increase (%)	Net Migration Rate	Population (millions)		Infant Mortality Rate[a]	Total Fertility Rate[b]	Percent of Population		Family Planning Among Married Women Ages 15-49[c]			Life Expectancy at Birth (Years)			Population Ages 15-49 With HIV/AIDS (%)		Urban Population (%)	Population per km² of Arable Land	GNI per Capita, PPP[e]	SPECIAL FOCUS COVID-19 INDICATORS: Excess Deaths Due to COVID-19 Pandemic, 2020-2021 Annual Average[f]			Population Fully Vaccinated Against COVID-19 (%)[g]
						mid-2035	mid-2050			Ages <15	Ages 65+	Using All Methods (%)	Using Modern Methods (%)	Demand Satisfied by Modern Methods (%)[d]	Total	Males	Females	Males	Females				No. of Excess Deaths	Excess Deaths per 100,000 Population	Excess Deaths as % of Total Deaths	
EUROPE	742	9	13	-0.3	2	740	724	4	1.5	16	19	66	56	—	78	75	81	—	—	75	555	42,820	1,365,595	182	13	68
EUROPEAN UNION	445	9	12	-0.3	3	445	439	3	1.5	15	21	65	—	—	80	78	83	—	—	75	613	48,451	566,819	128	10	74
SOUTHERN EUROPE	151	7	12	-0.4	2	148	144	3	1.3	14	22	63	51	—	81	79	84	0.4	0.1	72	695	39,273	215,838	140	11	76
Albania	2.8	10	11	-0.1	-12	2.7	2.4	10.0	1.3	16	16	46	4	6	77	75	80	<0.1	<0.1	63	463	15,430	6,345	221	15	43
Andorra	0.08	6	5	0.1	18	0.08	0.08	6.0	1.0	13	15	—	—	—	80	77	84	—	—	88	11,262	—	188	242	22	69
Bosnia-Herzegovina	3.4	7	14	-0.7	-8	3.2	2.9	4	1.1	15	18	46	12	22	77	75	80	—	—	49	339	16,690	7,858	240	17	26
Croatia	3.8	9	17	-0.7	0	3.7	3.4	4.0	1.5	14	22	—	—	—	78	75	81	0.1	<0.1	58	428	33,900	8,588	210	14	55
Greece	10.6	8	12	-0.4	1	10.1	9.5	3.2	1.3	14	23	—	—	—	81	79	84	0.3	<0.1	80	498	31,410	9,697	93	7	73
Italy	58.9	7	12	-0.5	3	56.8	53.7	2.8	1.3	13	24	65[h]	52[h]	—	82	80	85	0.3	0.1	71	862	46,490	80,401	133	11	79
Kosovo[i]	1.8	15	8	0.7	2	1.8	1.7	15	2.1	24	9	67	9	13	76	74	78	—	—	42	—	13,010	—	—	—	45
Malta	0.5	9	8	0.1	1	0.5	0.5	3.7	1.1	13	19	—	—	—	84	81	86	—	—	95	5,711	44,550	239	54	6	106
Montenegro	0.6	11	15	-0.3	-2	0.6	0.7	2.8	1.7	18	16	21	12	28	76	73	79	0.1	<0.1	68	6,801	23,280	1,955	311	21	45
North Macedonia	1.8	10	16	-0.5	-0	1.8	1.7	6	1.4	17	17	60	14	20	74	72	76	—	—	59	436	17,060	7,679	369	23	40
Portugal	10.3	8	12	-0.4	2	10.2	9.5	2.4	1.3	13	23	74[h]	67[h]	—	81	78	83	0.7	0.3	67	1,087	35,470	10,224	100	8	87
San Marino	0.03	6	9	-0.3	2	0.03	0.03	9.4	1.0	14	19	—	—	—	87	85	89	—	—	98	1,699	—	86	251	22	70
Serbia	6.8	9	20	-1.1	-0	6.3	5.7	4.7	1.5	14	21	62	21	30	73	70	76	<0.1	<0.1	57	260	20,610	27,822	319	19	38
Slovenia	2.1	9	11	-0.2	9	2.1	2.0	2.2	1.6	15	21	—	—	—	81	78	83	<0.1	<0.1	55	1,170	43,060	2,791	134	12	59
Spain	47.4	7	9	-0.2	3	48.3	49.9	2.5	1.2	14	20	62	60	—	83	80	86	0.5	<0.1	81	408	40,980	51,969	111	11	87

OCEANIA

	Population (millions) mid-2022	Births per 1,000 Population	Deaths per 1,000 Population	Rate of Natural Increase (%)	Net Migration Rate	Population (millions) mid-2035	Population (millions) mid-2050	Infant Mortality Rate[a]	Total Fertility Rate[b]	Percent of Population Ages <15	Percent of Population Ages 65+	Family Planning Among Married Women Ages 15-49[c]: Using All Methods (%)	Using Modern Methods (%)	Demand Satisfied by Modern Methods (%)[d]	Life Expectancy at Birth (Years): Total	Males	Females	Population Ages 15-49 With HIV/AIDS (%): Males	Females	Urban Population (%)	Population per km² of Arable Land	GNI per Capita, PPP[e]	SPECIAL FOCUS COVID-19 INDICATORS: Excess Deaths Due to COVID-19 Pandemic, 2020-2021 Annual Average[f]: No. of Excess Deaths	Excess Deaths per 100,000 Population	Excess Deaths as % of Total Deaths	Population Fully Vaccinated Against COVID-19 (%)[g]
OCEANIA	44	16	7	0.9	-0	55	66	15	2.1	23	13	59	56	—	78	76	81	0.3	0.3	67	1,258	39,522	-8,438	-20	-3	64
Australia	25.8	12	7	0.5	-0	33.2	40.6	3.2	1.7	19	17	67	65	—	83	81	85	0.2	<0.1	86	84	55,290	-7,128	-28	-4	85
Federated States of Micronesia	0.1	21	6	1.5	-6	0.1	0.1	30	2.9	30	6	—	—	—	70	69	72	—	—	23	5,299	3,900	-56	-48	-6	—
Fiji	0.9	19	9	1.0	-6	0.9	0.9	15	2.5	29	6	36	—	51	67	65	69	0.2	0.2	58	1,174	11,450	-59	-6	-1	71
French Polynesia	0.3	13	5	0.7	-1	0.3	0.3	5.4	1.7	21	10	—	—	—	77	75	79	—	—	62	11,234	—	—	—	—	—
Guam	0.2	17	7	1.0	-1	0.2	0.2	5.4	2.8	27	10	—	—	—	77	74	80	—	—	95	17,990	—	—	—	—	—
Kiribati	0.1	27	6	2.1	-2	0.1	0.2	38	3.3	36	4	34	27	54	67	66	69	—	—	56	6,137	4,050	-23	-19	-2	52
Marshall Islands	0.05	19	8	1.2	-44	0.05	0.05	24	2.7	32	5	45	42	81	65	64	67	—	—	78	2,722	5,120	-40	-67	-10	—
Nauru	0.01	28	7	2.1	-5	0.01	0.01	15	3.5	39	2	36	25	43	64	60	67	—	—	100	—	25,110	-1	-7	-4	—
New Caledonia	0.3	14	6	0.8	-4	0.3	0.3	11	2.0	22	11	—	—	—	79	75	83	—	—	72	4,595	—	—	—	—	—
New Zealand	5.1	11	7	0.5	-1	5.8	6.3	4.7	1.6	19	16	80[h]	75[h]	—	82	81	84	0.1	<0.1	87	975	45,440	-1,339	-28	-4	85
Palau	0.02	15	13	0.3	0	0.02	0.02	15	2.4	21	10	—	—	—	66	62	71	—	—	81	5,992	16,620	-20	-111	-11	—
Papua New Guinea	9.3	26	7	1.9	1	11.8	15.1	34	3.2	34	3	37	31	49	65	63	68	0.7	1.2	13	3,104	4,340	381	4	1	3
Samoa	0.2	29	5	2.3	-7	0.2	0.2	15	4.7	38	5	17	16	29	73	70	75	—	—	18	1,780	6,300	-45	-22	-4	101
Solomon Islands	0.7	30	5	2.5	-1	1.0	1.3	16	4.0	39	3	29	24	38	70	69	72	—	—	25	3,722	2,680	-30	-4	-1	27
Tonga	0.1	23	7	1.6	-9	0.1	0.09	10	3.2	34	6	29	25	49	71	68	74	—	—	23	496	7,260	-19	-17	-3	—
Tuvalu	0.01	25	10	1.5	-1	0.01	0.01	20	3.3	32	6	24	22	45	65	61	69	—	—	65	—	6,770	-6	-46	-6	—
Vanuatu	0.3	29	5	2.4	-1	0.4	0.5	20	3.7	39	4	49	37	51	70	68	73	—	—	26	1,540	3,220	-38	-12	-2	43

NOTES & SOURCES

Notes

- * The 14.9 million total excess deaths due to the COVID-19 pandemic in 2020 and 2021 are based on an annual average of 7.46 million.
- a Infant deaths per 1,000 live births. Rates shown with decimals indicate national statistics reported as completely registered, while those with no decimals are estimates from the sources cited on the reverse. Rates shown in italics are based on fewer than 50 annual infant deaths, so the figures are estimated from an average of the previous three years.
- b Lifetime births per woman (see Definitions on back for more information).
- c Data prior to 2016 are shown in italics.
- d Percent of married or in-union women with a need for family planning who are using modern methods.
- e In current international dollars. Data prior to 2021 are shown in italics.
- f Estimated deaths directly and indirectly associated with COVID-19 pandemic (see Definitions for more information). Negative values indicate the number of deaths averted due to lower risks of death from causes such as traffic accidents and influenza because of changes in people's behaviors.
- g Percent of the population that countries, territories, and regions report as being fully vaccinated as of mid-2022, according to their definition of the term.
- h Data refer to sexually active women, ever-married women, or all women. Data for Canada and Portugal refer to sexually active non-pregnant women and non-pregnant women, respectively.
- i Special Administrative Region.
- j Kosovo declared independence from Serbia on Feb. 17, 2008. Serbia has not recognized Kosovo's independence.
- (–) Indicates data unavailable or inapplicable.

The Data Sheet lists all geopolitical entities with populations of 150,000 or more and all members of the United Nations, including sovereign states, dependencies, overseas departments, and some territories whose status or boundaries may be undetermined or in dispute. **More-developed, less-developed**, and **least-developed** regions follow the UN classification (https://www.un.org/ohrlls/content/least-developed-countries). **High-income, middle-income** (composed of **upper middle-income** and **lower middle-income**), and **low-income** economies follow the World Bank classification based on GNI per capita (https://datahelpdesk.worldbank.org/knowledgebase/articles/906519-world-bank-country-and-lending-groups).

World and Regional Totals: Regional population totals are independently rounded and include small countries or areas not shown. Regional and world rates and percentages are weighted averages of countries for which data are available. Regional averages are shown when data or estimates are available for at least three-quarters of the region's population.

World Population Data Sheets from different years **should not be used as a time series.** Fluctuations in values from year to year often reflect revisions based on new data or estimates rather than actual changes in levels.

Sources

The rates and figures are primarily compiled from the following sources: online databases, reports, and other materials from national statistical offices and regional organizations; demographic and family planning surveys such as the Demographic and Health Surveys (DHS), Multiple Indicator Cluster Surveys (MICS), and Performance Monitoring for Action (PMA) Surveys; the UN *Demographic Yearbook 2020* and *Population and Vital Statistics Report* of the UN Statistics Division; *World Population Prospects: The 2022 Revision*, *World Contraceptive Use 2021*, and *World Urbanization Prospects: The 2018 Revision* of the UN Population Division; the International Data Base of the International Programs Center, U.S. Census Bureau; *World Development Indicators* online database of the World Bank; Global Excess Deaths Associated With COVID-19 (modelled estimates) data set of the World Health Organization; Johns Hopkins Coronavirus Resource Center of Johns Hopkins University & Medicine; AIDSinfo online database of UNAIDS; and FAOSTAT online database of the Food and Agricultural Organization of the United Nations. The sources also include direct communication with national statistical offices, demographers, and country experts from around the world. Specific data sources may be obtained by contacting the authors of the *2022 World Population Data Sheet*. Demographic rates for countries with complete vital registration are those most recently reported. For more-developed countries, the rates refer to 2021 or 2020. For other indicators, see Definitions for data years.

DEFINITIONS

POPULATION AND PROJECTED POPULATION, MID-2022, MID-2035, AND MID-2050
Current estimates and projections are based on recent censuses, official national data, or analyses conducted by national statistical offices, regional organizations, PRB, UN Population Division, or International Programs Center of the U.S. Census Bureau. The effects of refugee movements, large numbers of foreign workers, and population shifts due to current events are taken into account to the extent possible. Projected populations are based on reasonable assumptions on the future course of fertility, mortality, and migration.

BIRTHS/DEATHS PER 1,000 POPULATION
Crude birth and death rates, or the annual number of births and deaths per 1,000 total population, respectively.

RATE OF NATURAL INCREASE (%)
The birth rate minus the death rate, expressed as a percentage. This value represents the estimated rate of population growth without regard for migration.

NET MIGRATION RATE
The estimated net migration (immigration minus emigration) per 1,000 total population. For some countries, data are derived as a residual from estimated birth, death, and population growth rates. A value of "-0" indicates that the net migration rate is less than 0 but greater than -0.5; a value of "0" means that the net migration rate is greater than or equal to 0 and less than 0.5.

INFANT MORTALITY RATE
The annual number of deaths of infants under age 1 per 1,000 live births. Decimals indicate national statistics reported as completely registered; those without decimals are estimates from the sources cited above. Rates shown in italics are based on fewer than 50 annual infant deaths and, as a result, are subject to considerable yearly variability; rates shown for such countries are estimated from an average of the previous three years.

TOTAL FERTILITY RATE
The average number of children a woman would have assuming current age-specific birth rates remain constant throughout her childbearing years (usually considered to be ages 15 to 49).

POPULATION AGES <15 AND AGES 65+
The percentage of the total population in those age groups, which are often considered the "dependent ages."

FAMILY PLANNING AMONG MARRIED WOMEN (AGES 15-49): USING ALL METHODS, USING MODERN METHODS, DEMAND SATISFIED BY MODERN METHODS (%)
The percentage of currently married or in-union women (unless otherwise indicated in footnotes) of reproductive age who are currently using any form of contraception; any modern form of contraception; and among those with a need for family planning, the percentage who have their need met by modern methods. Modern methods comprise clinic and supply methods, including the pill, injectable, implant, IUD, condom, and sterilization. The inclusion of lactational amenorrhea and/or Standard Days Method in modern methods varies across countries. Data are from the most recently available national-level surveys since 2006. Data prior to 2016 are shown in italics. The classification of women as currently married or in-union may vary by country. The age range of women surveyed varies in some countries.

LIFE EXPECTANCY AT BIRTH, TOTAL AND BY SEX
The average number of years a newborn infant can expect to live under current mortality rates.

POPULATION AGES 15-49 WITH HIV/AIDS BY SEX (%), 2021
The percentage of the population ages 15-49 living with HIV/AIDS.

URBAN POPULATION (%)
The percentage of the total population living in areas termed "urban" by that country or by the UN.

POPULATION PER SQUARE KILOMETER OF ARABLE LAND
The mid-year 2022 population divided by the square kilometers of arable land last measured in 2020.

GNI PER CAPITA, PPP, 2021 ($ CURRENT INTERNATIONAL)
Gross national income in purchasing power parity (PPP) divided by mid-year population. Data prior to 2021 are in italics.

EXCESS DEATHS DUE TO COVID-19 PANDEMIC, 2020-2021 ANNUAL AVERAGE: NUMBER, RATE PER 100,000 POPULATION, AND EXCESS DEATHS AS A PERCENT OF TOTAL DEATHS
The difference between the average annual number of deaths in 2020 and 2021 and the deaths predicted for the same period if pre-pandemic trends had continued under normal circumstances as published by the World Health Organization. The difference reflects the number of deaths associated with the COVID-19 pandemic both directly through the coronavirus, SARS CoV-2, and indirectly through changes in access to health services and in people's behaviors. Negative values indicate the number or percent of deaths averted due to lower risks from certain causes of deaths, such as traffic accidents and influenza, due to changes in people's behaviors. Mean estimates are presented; low and high estimates are available from the WHO (https://www.who.int/data/sets/global-excess-deaths-associated-with-covid-19-modelled-estimates).

POPULATION FULLY VACCINATED AGAINST COVID-19 (%), MID-2022
The percentage of the total population fully vaccinated against COVID-19 as of June 30, 2022, based on the number of individuals that countries, territories, and regions report as being fully vaccinated. The definition of fully vaccinated may vary by country or region. Data were compiled by the Johns Hopkins Coronavirus Resource Center.

MIX
Papier aus verantwortungsvollen Quellen
Paper from responsible sources
FSC® C105338

If you have any concerns about our products,
you can contact us on
ProductSafety@springernature.com

In case Publisher is established outside the EU,
the EU authorized representative is:
Springer Nature Customer Service Center GmbH
Europaplatz 3, 69115 Heidelberg, Germany

Printed by Libri Plureos GmbH
in Hamburg, Germany